D1444720

# THE
# SPACE STATION

For Danny.

The continual adventurer...

Love,

Mum    12-25-97

# THE SPACE STATION

KENT ALEXANDER

GALLERY BOOKS
An Imprint of W. H. Smith Publishers Inc.
112 Madison Avenue
New York, New York 10016

**A FRIEDMAN GROUP BOOK**

Published by GALLERY BOOKS
An imprint of W.H. Smith Publishers, Inc.
112 Madison Avenue
New York, New York 10016

Copyright © 1988 by Michael Friedman Publishing Group, Inc.

All rights reserved. No part of this publication may be
reproduced, stored in a retrieval system or transmitted, in any
form or by any means, electronic, photocopying, recording, or
otherwise, without the prior written permission of the copyright owner.

ISBN 0-8317-7940-3

*THE SPACE STATION*
was prepared and produced by
Michael Friedman Publishing Group, Inc.
15 West 26th Street
New York, New York 10010

Editor: Sharon Kalman
Art Director: Mary Moriarty
Designer: Fran Waldmann/Rod Gonzalez
Photo Editor: Christopher Bain
Production Manager: Karen L. Greenberg
All photographs courtesy of NASA

Typeset by B.P.E. Graphics, Inc.
Color separations by South Sea International Press Ltd.
Printed and bound in Hong Kong by Leefung-Asco Printers Ltd.

# *Acknowledgments*

Thanks to: Wendy Baker for a good turn;
my mother; the NASA Photo Library in Washington, D.C.;
and all my friends for their support.

# Contents

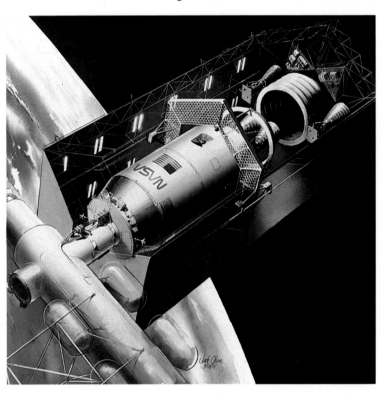

# CHAPTER 1

# The History of the NASA Program

In the early 1900s people read Jules Verne and H.G. Wells with an awe normally reserved solely for the gods. These writers sketched a world where space travel was routine, and exploration involved the search for life on other, distant planets. Yet, despite this romance with literature, the general public did not believe that space was the domain of mortals; only God belonged in the heavens.

Nevertheless, inventors like Robert Hutchings Goddard of the United States, Herman Oberth of Germany, and Konstantin Eduardovich Tsiolkovsky of Russia, all inspired by the same literature, theorized about the possibility of manned spaceflight—traveling through space in a rocket. Much of this theory was spawned by the new technology of World War II. Super rockets like the United States' *Titan, Redstone,* and *Atlas* required new work in aerodynamics, and the money channeled into the war effort assisted the postwar inventors in actualizing their dreams.

Russia, too, had begun research into spaceflight, and, in fact, it was the signal of the Russian satellite, *Sputnik,* on October 4, 1957 that launched the space race. On the heels of *Sputnik,* President Dwight D. Eisenhower signed the National Aeronautics and Space Act of 1958. In October of the same year, the National Aeronautics and Space Administration (NASA) was established to propel the United States into leadership in the space sciences.

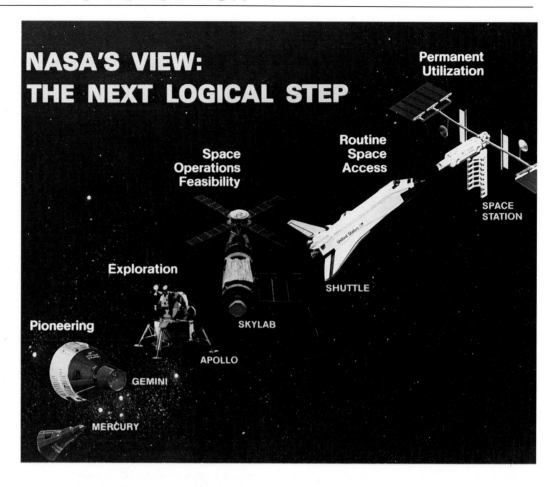

## NASA—The Early Years

From the outset, NASA officials decided that if they couldn't build rockets as large as the Russians were building, they would build the first rocket to send a man to the moon. With this goal in mind, NASA designed Project Mercury, a series of rockets built out of a ballistic missile coupled with the Air Force launch vehicle, *Redstone.* This first NASA rocket model built to send man into space was a far cry from the sophisticated space vehicles of today. In fact, it

was a fully automated system where the astronaut was not integral to the workings of the system, and was on board only to perform repairs if the system broke down. Serving as a "redundant component" to the vehicle was not what the original seven astronauts chosen for the first missions had in mind. Alan Shepard, Gus Grissom, John Glenn, Scott Carpenter, Walter Schirra, L. Gordon Cooper, and Donald Slayton vigorously objected to this concept of the role of the astronaut, stating that, at the very least, they should be able to override the automatic system of the rocket if it malfunctioned.

These soon-to-be heroes wanted more control, and their demands challenged NASA into a fresh realization—

space vehicles could and should be controlled by the astronaut. While subsequent missions into space began to incorporate the demands of the astronauts, the first Mercury rocket was truly an astronaut's nightmare. In order to keep the weight of the vehicle down, all the dimensions of the craft were extremely tight. The majority of the room inside the capsule was filled with instrument panels, wires, hookups, a radio, and an emergency parachute. In the midst of this claustrophobic array was the astronaut; his seat was sculpted to his back and legs in order for him to fit. Because the capsule designers believed a window would probably rupture from the change in atmospheric pressure, two tiny portholes were constructed above

**L**eft: In 1982, NASA began to plan and define a possible space station. With the completion of the space shuttle flight tests, the station, utilizing the nation's past space experience, is seen by NASA as the next logical step in space. A modified *Redstone* booster rocket (right) is poised and ready to lift a production version Project Mercury spacecraft from Cape Canaveral, Florida. Project Mercury was NASA's first manned space flight program. The purpose of this test flight was to study the effects of a period of approximately five minutes of weightlessness.

Left: November 29, 1961; 10:07 EST. The Project Mercury primate-carrying spacecraft is launched from Cape Canaveral, Florida in a test designed to qualify the spacecraft, and its orbit and reentry systems. An *Atlas* launch vehicle built by General Dynamics/Astronautics was used to launch the craft, which was manufactured by McDonnell Aircraft Corporation. Right: On August 21, 1965 NASA launched *Gemini V* on an eight-day orbital mission. Astronaut L. Gordon Cooper was the Command pilot, with Astronaut Charles Conrad as pilot. This mission would be the longest manned space flight to date.

the astronaut's head. A periscope was his sole method of viewing the outer world. Wisely, after this first flight, the capsule's design was altered to allow for a pilot's window, as well as a hatch to be opened by the pilot after landing.

Before Shepard, the first man to venture into space, was launched, NASA sent several unmanned rockets, mice, and chimpanzees into the air. "Monkeynauts" were specially trained for the *Mercury Redstone 2* flights, which tested the environmental control and recovery systems.

# To The Moon

In December of 1960, the first Mercury Redstone was launched into suborbital flight. On May 5, 1961, Commander Alan Shepard, Jr. was launched in the first of six Mercury spacecraft by the launch vehicle, *Mercury 3*. The command module, *Freedom 7,* completed a fifteen minute and twenty-two second suborbital flight at an altitude of 116 miles (185.6 kilometers). During this short flight, Shepard traveled 297 miles (475.2 kilometers), demonstrating how to manually control the spacecraft while weightless. From the third launch on, the *Atlas D* rocket was used instead of the *Mercury 3* for all Mercury flights. This launch rocket had much more thrust, or power, and propelled John Glenn, the first American to orbit the planet, through three Earth revolutions on February 20, 1962. L. Gordon Cooper, making a total of twenty-two orbits around the earth in just twenty-four hours, was the last astronaut in the

Project Mercury series. During this year, 1962, NASA also launched *Telestar 1,* the first privately owned (American Telephone & Telegraph, or AT&T) communications satellite. The following month, NASA launched *Mariner 2,* a Venus probe craft.

Throughout the early 1960s NASA grappled with the mandate of President John F. Kennedy: to put an American on the Moon before the Russians did; however, there was still much to be learned about spaceflight before an astronaut could be hurled toward that destiny.

Following Project Mercury was the Gemini series, a manned spaceflight project designed to answer many of the questions of space rendezvous and the mechanics of extravehicular activity, both crucial components of the planned Moon landing. The Gemini design was identical to that of the Mercury capsule except that the interior was enlarged to accommodate a two-man crew. Using the Air Force launch vehicle *Titan II* to increase thrust, the Gemini command module, while still in Earth orbit, allowed astronauts to disembark from the capsule and perform numerous extravehicular maneuvers in space while attached to a 25-foot (8-meter) umbilical cord. Astronaut Edward H. White II was the first American to actually leave a spacecraft, the *Gemini 4,* while in orbit. White remained outside the craft for twenty-one minutes, using a handheld Self Maneuvering Unit (a small gas gun) to move about in the zero gravity environment of space. While White cavorted about in space, the command pilot, James A. McDivitt performed other scientific and engineering chores inside the craft. Four months later, *Gemini 7* succesfully rendezvoused in orbit with *Gemini 6.* Soon after, NASA achieved the first docking between a manned

spacecraft, the *Gemini 8,* and an unmanned vehicle, the *Agena.*

While the Gemini program was being tested in space, NASA, using the more powerful *Saturn* launching rocket, had begun the *Apollo* project, which tested the ability of space vehicles to traverse gravitational fields. On July 16, 1969, after ten *Apollo* missions, *Apollo 11* lifted off for the Moon. Astronaut Neil Armstrong announced upon landing on the Moon: "Houston . . . Tranquillity Base here . . . the Eagle has landed." Television viewers all over the world watched Armstrong set foot on the alien lunar landscape and declare, "One small step for man; one giant leap for mankind." While astronaut Michael Collins sat in the *Columbia* command module circling the Moon, Edwin E. Aldrin, Jr. joined Armstrong, roaming the lunar landscape, collecting samples of Moon rock and soil. President Kennedy's mandate had been met with resounding success!

Ensuing *Apollo* missions explored much more of the lunar landscape, employing roving vehicles and conducting scientific experiments. When finished, the *Apollo* project had put twelve men on the Moon in less than twelve years.

Left: The April 8, 1964 *Gemini/Titan I* launch at Cape Kennedy Missile Test Area. Right: Astronaut John W. Young, commander of the *Apollo 16* lunar landing mission, leaps from the lunar surface as he salutes the United States flag during the first *Apollo 16* extravehicular expedition.

An artist's 1978 conception of a free-flying power station (right) to support long-duration space shuttle flights of up to 120 days. The astronaut performing an extravehicular activity (EVA) provides a size comparison to the power station. The shuttle orbiter is seen below with its cargo bay doors open. Far right: The reusable spaceship *Columbia* begins its third trip. On board are Commander Jack Lousma and Pilot Gordon Fullerton. Aloft on twin clouds of smoke and fire, *Columbia* carried an OSS-1 pallet of experiments for materials processing; two for studying plant growth in zero-gravity; a "Getaway Special" test canister; and the Remote Manipulator System.

## The First American Space Station

During this time of space heroes walking on the Moon, American enthusiasm for the space program was boundless. Films like *2001: A Space Odyssey* were box office smashes. NASA began to anticipate an age when space exploration, with the support of larger fiscal budgets, would flourish. In 1969, NASA launched *Mariner 6*, a fly-by probe of Mars, and in January 1972, President Richard Nixon announced approval of the Space Shuttle program. Within this context, the first American manned space station, *Skylab*, was born. *Skylab's* principle objective was to determine if people could physically withstand arduous stays in space as well as perform necessary tasks. However, the *Skylab* mission was also important due to the valuable information gathered about the Sun, made possible by equipping the space station with a special solar telescope.

Built from the S-IVB stage of a *Saturn V* Moon rocket, *Skylab* was the pioneer space vehicle for living within the confines of space. A converted hydrogen tank was utilized to house the three-man crew; 13,000 cubic feet (368 cubic meters) were used for meals, showers, heating, food storage, and a large compartment for clothing, towels, and toiletries. Due to budget restraints, *Skylab* was cut back to only one space orbital workshop on May 14, 1973, and three subsequent astronaut flights. The knowledge gained from *Skylab*, however, was priceless. It was discovered that people could indeed live and work in space without incurring physiological problems. Industry also realized that

certain processes were better done in space. On July 17, 1975, before NASA launched *Viking 1* and *Viking 2* towards Mars, another maneuver was accomplished in space when an Apollo spacecraft linked up with a Soviet Soyuz in orbit. One hundred and forty miles above the Earth, the Apollo-Soyuz Test Project was the first international manned spaceflight. Three United States astronauts, Tom Stafford, Donald "Deke" Slayton, and Vance Brand, along with the Soviet cosmonauts, Aleksey Leonov, and Valerey Kubasov, shared meals and jokes, conducted science experiments, and exchanged presents. This project was the result of six years of joint negotiations, illuminating the path for future cooperation between the two superpowers.

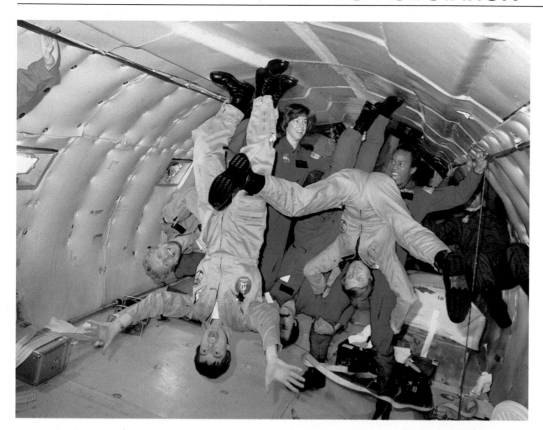

In the foreground, the crew of the *Spacelab D-1,* Messerschmid, Merbold, and Furrer, clown around. Bottom: At the Kennedy Space Center in Florida, the orbiter *Columbia* is backed into the Vehicle Assembly Building after being rolled out of the Orbiter Processing Facility.

Passing behind the Apollo *Saturn V* is the space shuttle *Discovery*, en route to the Vehicle Assembly Building.

# The Enterprise Becomes Reality

In the early 1970s NASA began to plan an improved and vastly superior space machine, the Space Shuttle Transportation System. After approval from President Nixon and Congress, the Space Shuttle became NASA's top priority. The Space Administration saw the Shuttle program as the answer to a myriad of questions—most importantly, the problem of expendable launch vehicles being inefficient and costly, and the ultimate goal of a permanent space station. Today, of the Shuttle's three main components, the space plane (the orbiter), two rocket boosters, and the external tank, only the latter is not reusable.

On August 12, 1977 the first orbiter, the *Enterprise*, made its maiden voyage with the assistance of a Boeing 747. Astronauts Fred Haise, Jr. and Charles Fullerton were propelled to an elevation of 22,800 feet (6,949 meters) before making the first free-flight in space. The Space Shuttle *Enterprise* made a total of five flights before NASA launched the Space Shuttle *Columbia* in 1981, heralding the demise of expendable space vehicles.

Four orbiters, *Columbia, Discovery, Atlantis,* and *Challenger,* were created for the new project. The vehicles are the size of a commercial airliner, 122.2 feet (37 meters) long, with a delta-shaped wing 78.06 feet (24 meters) across, a vertical fin, and a rudder with adjustable flaps to serve as air brakes. Inside are two levels: the upper level containing the flight deck, able to house a crew of seven, and the lower deck, the living area. Behind this is an enormous depressurized area—the cargo bay—where industrial or scientific cargo can be hauled into or back from space.

The orbiter is mounted on the side of an external tank to which the two solid rocket boosters are strapped. At lift-off, the orbiter's liquid-fuel engines are fired together with the rocket boosters. After clearing the launch tower, the Shuttle accelerates to the speed of sound and minutes later, the solid rocket booster's fuel is exhausted and the rockets are blown free from the external tank of the vehicle. These boosters, which descend via parachute into the ocean, are retrieved by waiting ships for reuse.

The Shuttle continues gaining speed and altitude until reaching a height of 75 miles (120 kilometers). At this point, the empty external tank is jettisoned from the orbiter. The orbiter's main engines are then cut off and the twin Orbiter Maneuvering System (OMS) engines are fired, boosting its speed to 17,500 miles (28,000 kilometers) per hour. At an altitude of 150 to 200 miles (240 to 320 kilometers), it begins its one and one-half hour orbit around the earth.

When the mission is complete, the

orbiter brakes, allowing the Earth's gravity to take hold of it. Turning tail first, the OMS rockets are fired against the direction of the flight and the vehicle descends, glider-like, until it reenters the atmosphere where, traveling at twenty-five times the speed of sound, it begins to rotate and lose speed. At 1,700 feet (518 meters) the orbiter's nose tilts up and the vehicle begins the descent that will eventually permit a horizontal landing on the three-mile runway at Kennedy Space Center in Florida.

# The Next Step

The search for new space goals began in earnest after the tragic *Challenger* accident on January 28, 1986. Suddenly NASA was under public scrutiny and was faulted for its poor management of the Shuttle program. This was viewed to be only a part of an unclear and incoherent space policy. NASA, according to its critics, had a transportation system but little else, as it had focused total attention on the Shuttle program. The triumphs of the Moon and the exploration of the solar system seemed like distant memories.

Once again the Soviet Union, with the successful launching of its space station, *Mir,* in 1986, played a part in the dilemma. An independent NASA advisory council announced in early 1987 that American leadership in space "absolutely requires the expansion of human life beyond Earth" and urged the exploration of Mars as the agency's primary goal. Dr. James C. Fletcher, NASA's administrator, appointed the first female astronaut, Sally K. Ride, to direct a study of what NASA's goals

MSFC-76-PA-4000-501

**A**n artist's rendering demonstrates the three necessary steps leading to the development of an assembly and fabrication facility in space. The first flight would involve producing several structural members by an automated beam fabrication module in the space shuttle cargo bay. The second step involves beam fabrication, plus limited erection and assembly, while the third would be fabrication, erection, and assembly of a large structure, such as the 100 kw solar power facility.

The view of Mars from the *Viking* spacecraft as it approaches the planet.

should be into the twenty-first century.

The findings of this study placed four primary plans in the forefront: the first is exploration of Mars. The other proposed initiatives involve an expanded study of Earth from orbit, an accelerated program of solar system exploration by unmanned spacecraft, and the establishment of a permanent scientific base on the Moon. Of these four "leadership initiatives," the Mars goal is seen as the most exciting and challenging. But whatever plan is chosen, and certainly if the Mars initiative is accepted, the space station is crucial.

NASA believes that once the Space Shuttle is again operative, a space station is the next logical step towards future ventures to the Moon, flights to Mars, and further Earth and solar system research, as well as zero-gravity processing of materials for industry. The development of such a space station will increase the manufacturing and technological capabilities of the United States, assuring its continued leadership in space through the 1990s and the years to come.

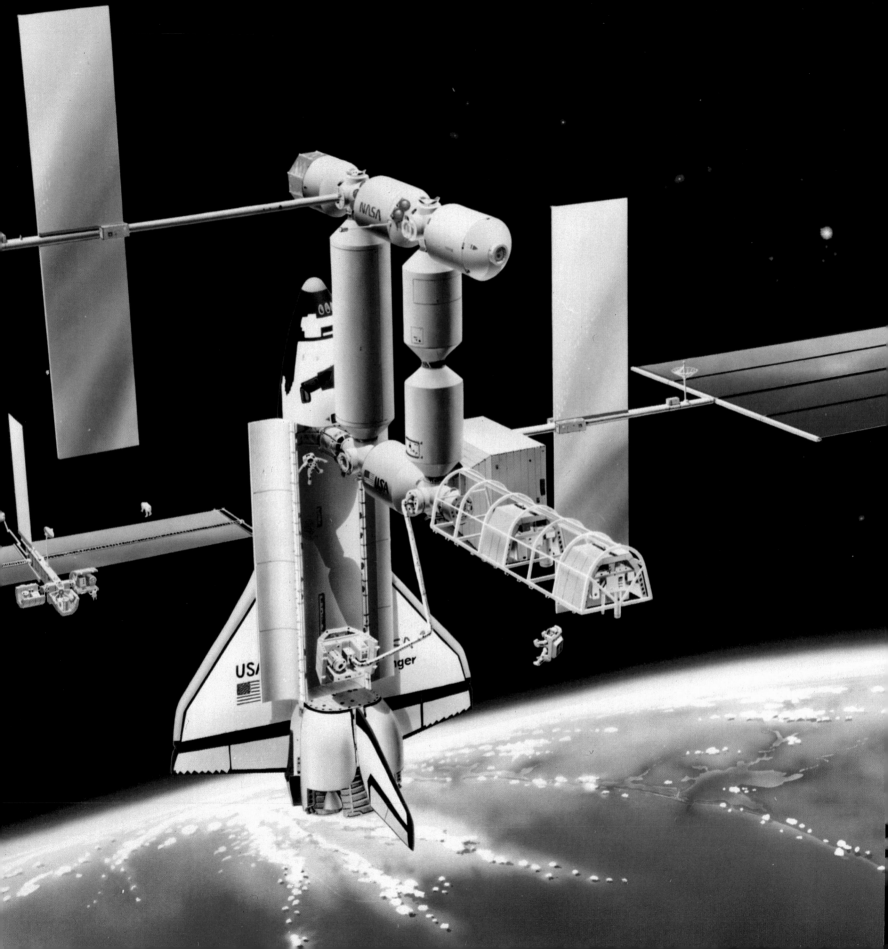

# The Birth of the Idea

In his January 1984 State of the Union address, President Reagan issued a directive to NASA to develop a permanently manned space station by the late 1990s. However, three years earlier, NASA had formed the Space Station Technology Steering Committee to assess the technological needs for a space station, and to produce a project definition called for by the Hearth Report of 1981. In 1982, under

This artist's rendering illustrates a base structure consisting of modules clustered around the large, wing-like solar panels. The modules would provide living quarters for up to eight people, lab and working areas, storage, utility, docking, and transfer facilities. The rack-like structure would provide room for mounting experiments designed to operate in space rather than in the pressurized modules. A space shuttle is seen docked during a mission to resupply the station or rotate the crew. At the upper right, a communications satellite is attached to an orbital transfer vehicle for transfer to geosynchronous orbit.

Previous page: This illustration by NASA's Lyndon B. Johnson Space Center in Houston, Texas shows one space station design configuration. Right: The Space Station Task Force has focused its activities on mission requirements, technology options, and preliminary systems engineering. Shown here is one possible space station configuration developed by NASA's Johnson Space Center in Houston, Texas.

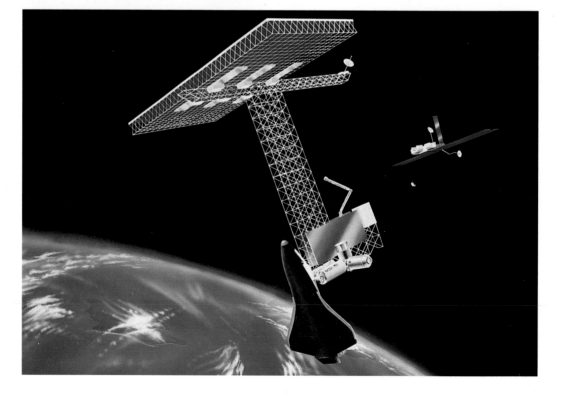

The space station module pictured here was designed by the Concept Development Group of the NASA Space Station Task Force (SSTF) and built by the NASA Langley Research Center in Hampton, Virginia. This was the model used by NASA Administrator James M. Beggs in his presentation to President Reagan in December, 1983. The President then presented this model at the London Economic Summit in June, 1984.

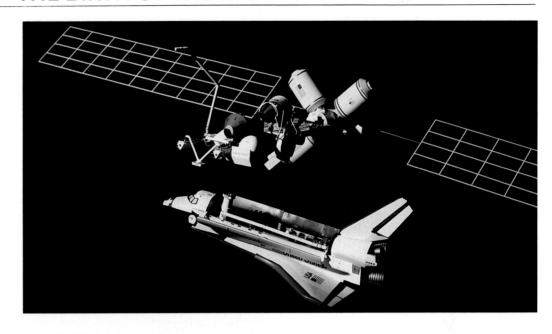

A possible space station configuration (left) developed by NASA's Johnson Space Center.

the protection of the Space Station Task Force (SSTF) this amalgam of engineering, scientific, financial, and administrative experts from each section of the agency conducted a two year study to clarify the goals for the proposed Space Station. According to NASA, these goals are:

To establish the means for a permanent and productive presence of people in space

To establish routine, continuous, and efficient use of space for science, applications, technology development, and operations

To further develop the commercial use of space

To develop and exploit the synergistic effects of the man-machine combination in space

To provide essential system elements and operations practices for an integrated, continuing national space capability

To stimulate the mutual benefits traditionally derived from cooperation in space with allies and friends

To reduce the cost and complexity of living in and using space

To be a major contributor to United States leadership in space in the 1990s and beyond

To ensure that the elements of the Space Station are compatible with, and can interface with, space elements of the operational Space Transportation System

To motivate future scientists and technologists, and provide leadership in furthering their education

An illustration of another possible space station design by NASA's Johnson Space Center.

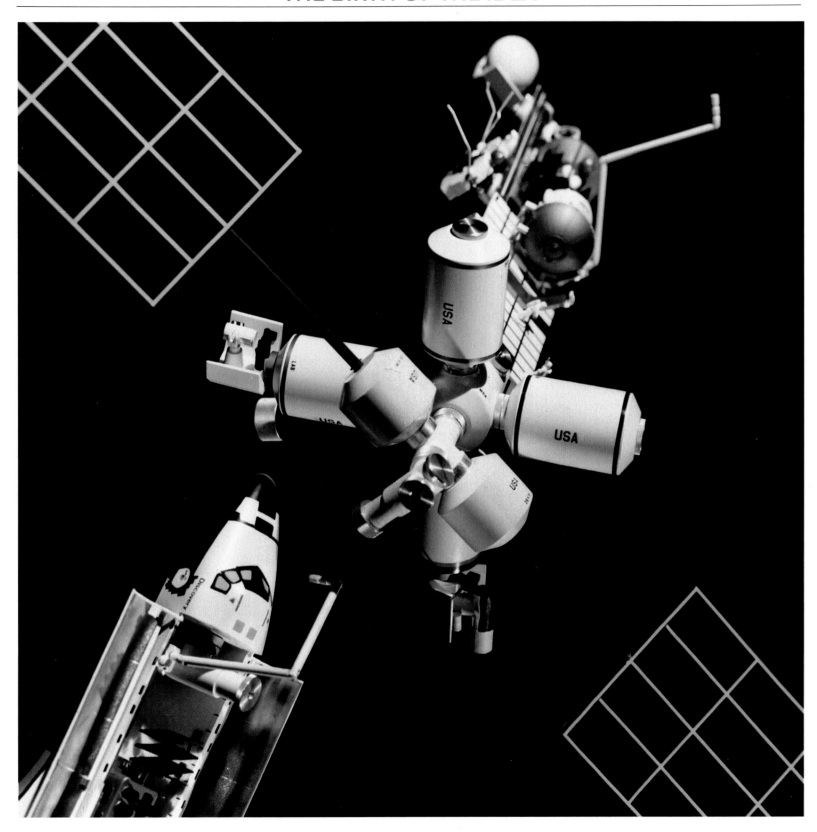

Steve Irick, of the Concept Development Group of the SSTF, is photographed here with the space station model used by Administrator Beggs in his presentation to President Reagan in December, 1983.

The baseline configuration (the most basic structure of the Station) outlined included a manned module at an orbit of 28.5° and two platforms, one at low orbit inclination and one in polar orbit.

The SSTF obviously saw the Space Station as an essential part of maintaining United States leadership in space activities, and the utilization of the economic and scientific opportunities offered by space. In 1984, the SSTF was replaced by the Interim Program Office and assigned responsibility for NASA-wide program development. By the end of 1987, three organizations, one civil service and two outside, were set up in the Washington area, attempting to meet the President's deadline for an operative space station. Most of the nearly five hundred managers were relocated to the Washington area. Because of the Station's program requirements for long lifetime, evolutionary growth, and inter-

**A**bove: This illustration is the space station reference configuration prepared by the TRW Space and Technology Group of Redondo Beach, CA. Right: An artist's rendering of one possible station concept. This concept was derived from earlier space platform studies carried out for NASA by the TRW Space and Technology Group. Two solar panels extend outward from the space craft. Extending upward is a large single radiator. An orbital transfer vehicle is parked to the rear of the station.

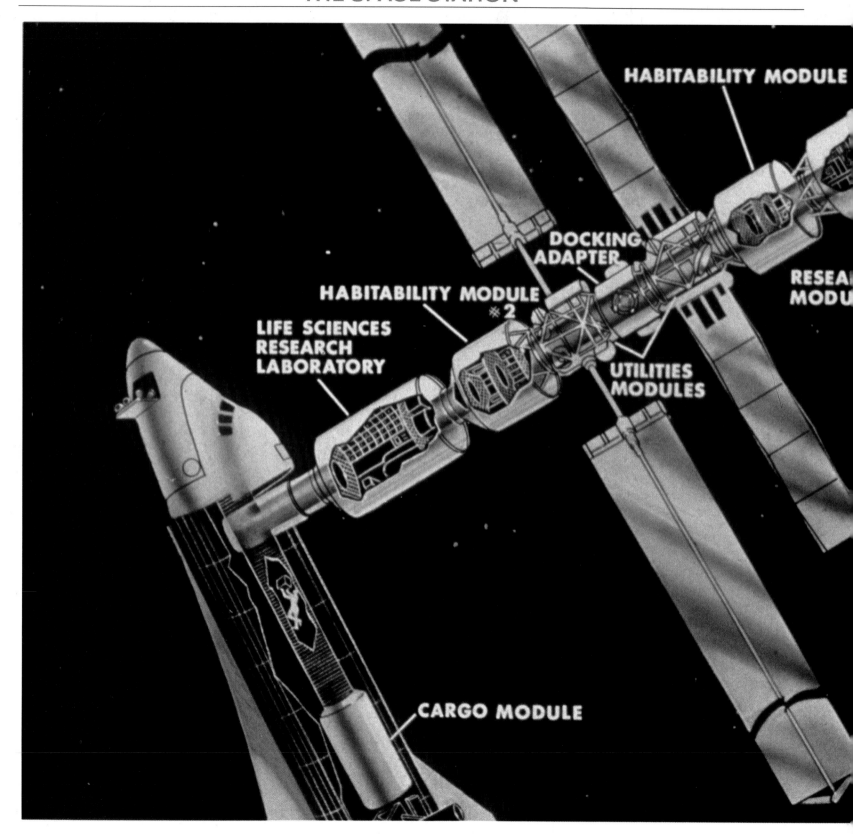

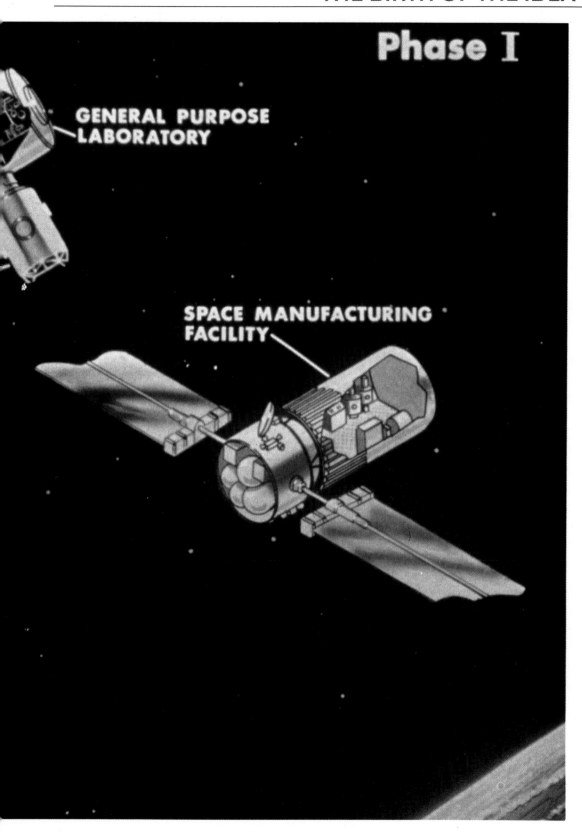

## Phase I

GENERAL PURPOSE LABORATORY

SPACE MANUFACTURING FACILITY

**S**PACE FACILITY EVOLUTION— studies in ''minimum''-type concepts of a space station utilizing *Skylab* and *Spacelab* technology.

**A**bove: A space station platform reference conception by McDonnell Douglas Astronautics Company of Huntington Beach, California. This particular station is composed of rotating pallets capable of holding science and applications payloads and an airlock joining two manned modules and a logistics module. Below: A possible space station conceived by Rockwell International's North American Space Operations Division. The artist's conception also shows an Orbital Transfer Vehicle (OTV) returning to the station after delivering a payload to higher orbit. Space-based at the station, such a vehicle would be both more economical and have greater space capabilities than an Earth-based vehicle. Pictured here (right) is the first element of the space station to be placed in orbit—the "utility section"—consisting of power, communications, thermal, and other core systems for the station.

national cooperation, NASA is contracting two important portions of the project to outside organizations: program requirements and assessment, and systems engineering and integration. NASA will utilize the data and analyses developed by its own personnel, the contractors, and potential international participants in choosing the initial and subsequent Space Station designs. A major objective of the plans is to define the proper mixture of human and nonhuman (machine) functions in the Station, setting an order for tasks to be properly distributed. NASA's ultimate desire is for machines and humans to augment and complement each other's abilities.

This management's structure of contracting to outside organizations parallels the earlier structure of the Apollo program of the 1960s. With the Apollo program, NASA consigned these functions to AT&T, which received the contract to build a subsidiary, called Belkom, to oversee the technical requirements for the Moon landing; and to Boeing, which managed all the scheduling and integration through the Technical Integration Effort (TIE). Similarly, the hardware exclusion clauses were contracted to outside organizations.

The systems engineering contract, the first to be awarded, was bid upon by six major aerospace companies: the Boeing Company, the Martin Marietta Corporation, the McDonnell Douglas Corporation, the General Electric Company, the Rockwell International Corporation, and as a separate entity, the Rocketdyne division of Rockwell International. The successful bidder will serve as a prime contractor on the Space Station. Because of the hardware exclusion contract clause, however, the company must not bid on the work pack-

This painting depicts a shuttle orbiter vehicle visiting a Space Operations Center in Earth orbit. This low-Earth orbit station is considered extremely important as a staging station for supporting numerous programs.

ages (see chapter 3).

Both the systems engineering contract and the four hardware elements or "work packages" contracts will be selected by NASA through requests for proposals. Each contractor has been asked to submit two proposals: one to cover the initial design, and the other for the construction of a fully enhanced station.

# How Much Is That Station in the Window?

The Space Station for which NASA is currently seeking bids differs considerably from the 1984 version. It is less ambitious, smaller, and probably more controversial. Several reasons exist for this. The *Challenger* disaster of January 1986 sent the entire space industry into shock. The destruction of this Shuttle, one of four, has greatly affected each program that depends, as the Station does, on numerous shuttle flights.

Another major factor in NASA's decision to scale-down the Station is its increasing cost. The original plan NASA

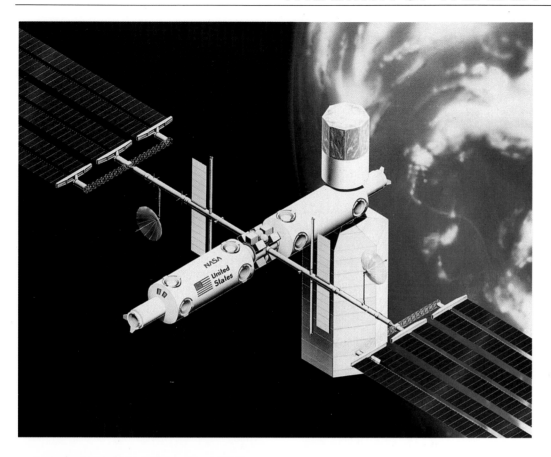

**A**bove: This artist's drawing shows an enclosed space garage (lower right) docked to the second command module. A logistics vehicle (upper right) is docked to the command structure. Left: A reference drawing produced by McDonnell Douglas Astronautics Company of Huntington Beach, California.

visualized was a Station estimated at a total of $8 billion. Agency officials soon revised this estimate to $13 billion, along with another $1.5 billion for engineering ground support. The present design is a compromise; this "barebones" Station will only require the $8 billion initially proposed. NASA has, in turn, requested bidding contractors to present ideas for lowering their costs. NASA officials believe that erecting this more modest version will cut costs and that the Station should be gradually developed over the coming years. This present attitude is closer in strategy to that of many prominent space scientists who fear a more expensive Station would be a financial burden and would only serve to threaten investment in other space projects.

It is possible that some of the financial burden will be absorbed through international participation. In the same State of the Union address in which he called for a permanently manned space station, President Reagan also invited allies of the United States to participate in the development and use of such a station. In keeping with Reagan's offer, the fourteen-nation European Space Agency (ESA) has agreed to supply a $2 billion laboratory to be called *Columbus*, and Japan has pledged a $1 billion labo-

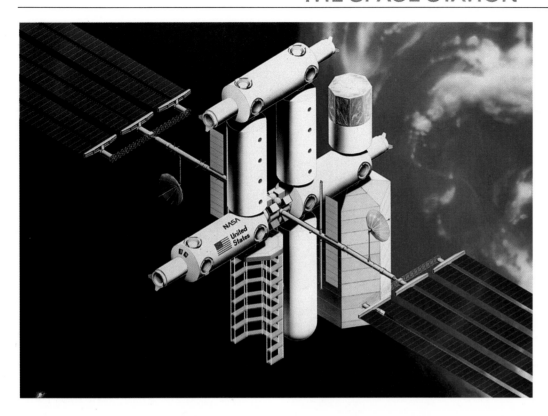

An artist's concept of the Space Operations Center, permanently manned with a crew of eight or twelve, orbiting Earth. Seven or eight Shuttle launches would be needed to bring the center to this stage of completion.

ratory to be attached to the central station. Both agreements, however, depend upon the current negotiations among the governments involved.

Another factor which has put NASA in an extremely difficult position is the Reagan administration's new emphasis on early deployment of the Strategic Defense Initiative (SDI), commonly called the "Star Wars" plan. NASA's administrator, Dr. James C. Fletcher, has said that the United States has no intention of placing weapons systems on the Space Station, even though the Pentagon has been granted permission to utilize the Station for non-specified research. NASA needs, and desires, international participation to alleviate some of the financial burden, as well as to establish a framework for future cooperative ventures; however, the agency cannot afford to desert such a powerful potential user as the Pentagon.

# A Stop On the Highway to Space

With the exception of the Space Shuttle, NASA's past programs have not been designed with evolutionary capability. A "permanently manned Space Station" must, by definition, span decades and be capable of accommodating evolutionary growth through slow addition, modification, and replacement. To guarantee that its useful life is not short, the Space Station must have the ability to evolve in capacity, capability, and technology. NASA believes that a properly designed Station will provide the necessary groundwork for succeeding space projects, whether they are manned operations in or beyond the Earth's orbit,

This artist's conception (right) speculates on a base structure composed of a number of modules clustered around a triangular beam structure, upon which solar cells could be installed to provide electrical energy to the station. A space shuttle is shown docked to the station.

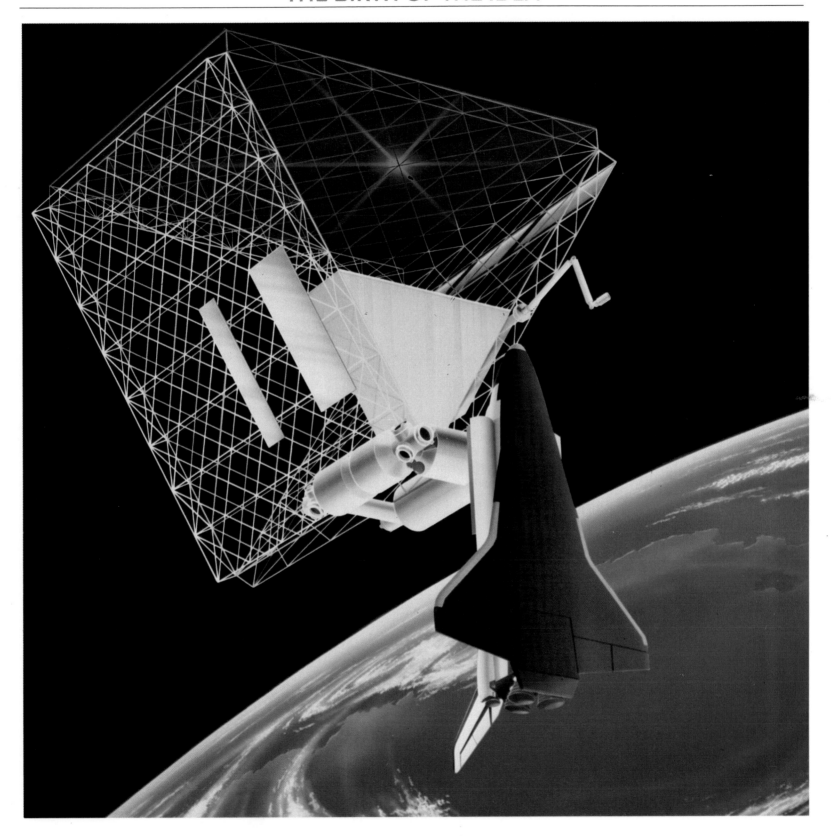

lunar-based operations, the mining of precious resources from the Moon or neighboring asteroids, or sample returns from missions to Mars. Also, the Station, with its ability to perform extensive and comprehensive space servicing, would ideally be able to maintain and update space telescopes in orbit, eliminating the need to return them to Earth for maintenance. Experiments on *Skylab* will become lengthier and will be modified so that they can be conducted on or near the Space Station. While these experiments will depend upon the Station for servicing, assembly, and manned interactions, so will experiments like the Cosmic Ray Measurements. This series of tests measures particles of very high energy and low flux, investigates their charge composition, energy spectra, and the isotopic composition of the cosmic ray nuclei. Also dependant upon the Space Station is the Pinhole Occulter Facility, which will be a system for imaging hard x-rays to provide high-resolution studies of x-ray bright points, active regions, solar flares, and the nature and structure of the Sun's corona (the faintly colored luminous ring that can be seen through a haze or thin cloud).

The Space Station will also benefit the launch and recovery phases of future solar system exploration missions. Modular launch systems, used to launch the separate pieces of the module, could be assembled easily at the Station to obtain better gravitational escape performance. Recovery of returned samples would provide a means for zero-gravity laboratory studies and provide major assistance in areas such as quarantine and thermal and physical protection of samples during reentry to the Earth's orbit.

The Station should also provide the first opportunity for extended work with animals, plants, and experimental equip-

**A**t NASA's Langley Research Center, researchers use a mobile work station to construct a beam from graphite-epoxy conical tubes. This would allow a pair of pressure-suited astronauts to assemble large structures in space. When a section of the structure is completed, it is moved along the assembly line; its size is extended by adding more structural elements or equipment. The experimental model will enable researchers to uncover difficulties that might be encountered in space and to identify assembly aids to improve productivity.

ment in weightlessness. Plans to capitalize on that opportunity include the development of a module to support studies of changes in physiological functions, such as calcium excretion and adaptation to increasing lengths of exposure to weightlessness. This module habitat will also allow the study of the use of artificial gravity and of the possible effects of weightlessness.

Because of these long-term needs, the Station must not only be designed for in-orbit maintenance but also be able to operate independently of the ground as much as possible. Ideally, NASA sees the Station as completely autonomous, except for resupply of personnel and materials. It is intended that there be no need for full-time monitoring of the usual subsystem status, detailed crew work scheduling, or crew health check assessments by teams of engineers, scientists, and Earth-based mission controllers. Most of the Station's routine monitoring tasks would ideally be done by computer. As Dr. Fletcher said in a speech to the space community, " . . . a Space Station is the next step in NASA's, and this country's, future plans. The Space Station will be the first step in a highway to space."

P.J. Weisgerber 81

# *The Work Packages*

In April 1983, the Space Station Task Force managed and coordinated trade-off studies, conducted preliminary cost analyses, and integrated the results of the mission requirement studies into a set of functional capabilities, or the "architecture" for the Space Station. The first stage of this plan was completed in May 1983 in concert with a study conducted by the Senior Interagency Group at

Previous page: Here, TRW's space platform concept is developed to one of its potentials as a manned structure. This particular station concept is composed of two solar arrays, a space radiator, three rotating pallets, and an airlock joining two manned modules.

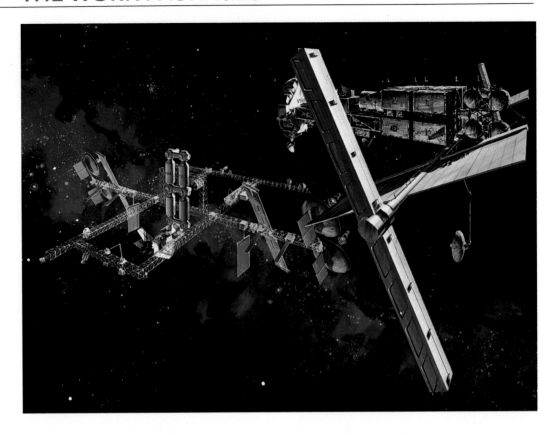

Left: One of TRW's unmanned space platform concepts shown from an underside view. Right: In this conception by TRW, a space platform is tugged toward the space station by an OMV.

the request of President Reagan.

The first phase of the Space Station is designed to provide a permanent manned presence in space at an orbital slant of 28.5°. The Station as planned is based on a 135 meter horizontal boom, or beam, made of an advanced fiber-glass-like composite material. The hub of the Station will be formed by four pressurized modules. Two of these will be provided by the United States and, if international negotiations continue to be successful, one module will be provided by the European Space Agency and one by Japan. One of the United States modules will permit astronauts to perform experiments in microgravity and in the life sciences: the other will be for sleeping, relaxing, and dining.

Each end of the boom will house photovoltaic (chemically produced electricity) arrays which will deliver 50 kilo-watts of power from conventional solar cells. This will be the Station's primary source of electrical power. The Station will also include a Canadian-built Mobile Service Center. This remote control "servicing arm" will perform basic maintenance work on the Station. Also included in the plan is a free-flying, unmanned polar-orbiting platform with remote sensing instruments. This platform will make valuable Earth and solar observations. Eight astronauts will be housed permanently on the Station and, while the original plan called for a ninety-day tour of duty for early crews, the newest plan has extended duty tours up to four months. After the first year of manned operation, NASA intends to increase the crew duration time to six months. Space Shuttles ferrying crews and supplies would dock at the habitation modules.

Phase two, to be implemented in the late 1990s, will add vertical trusses to the existing horizontal boom, which will hold additional research instruments (see page 86). The new additional upper section will anchor astronomical instruments while the lower one will hold instruments for observing Earth, as well as orbiting maneuvering vehicles, a shed-like facility for servicing free-flying commercial and noncommercial satellites, and a co-orbiting platform housing sensitive instruments that need to be left undisturbed. Mammoth wings of silicon attached to this truss would not only boost the Station's power to 87.5 kilo-watts by utilizing additional solar power, but also would refocus some of the energy as microwaves to be beamed to the Earth or elsewhere.

Beginning in 1994, NASA plans to utilize sixteen Space Shuttle flights to undertake the first phase of developing the Station. By late 1995, according to the plan, permanent operational capability will be finished. The first two Shuttle flights will carry the functional framework into orbit and assemble it. This framework includes the boom and the power system (see chapter 7). The third and fourth flights will carry the payloads. Once the polar platform is launched, the United States laboratory module will be orbited and assembled, though in bare-bones form because of its tremendous weight. Following this, the living quarters will be added, and, by the eleventh flight, NASA believes four astronauts will inhabit the Space Station. Once the photovoltaic arrays can deliver at least 75 kilowatts of power, the European and Japanese modules will be orbited. Completing the assembly will be extra equipment and the experimental setups containing permanent living quarters for the crew. After the sixteenth shuttle flight, all flights will carry new supplies and any additional work for the Space Station.

# THE WORK PACKAGES

Left: This rendering by Rockwell International of Downey, California depicts a growth phase with four U.S. modules, two lab modules, two habitation modules, a logistics module, and the European Space Agency and Japanese experiment modules.

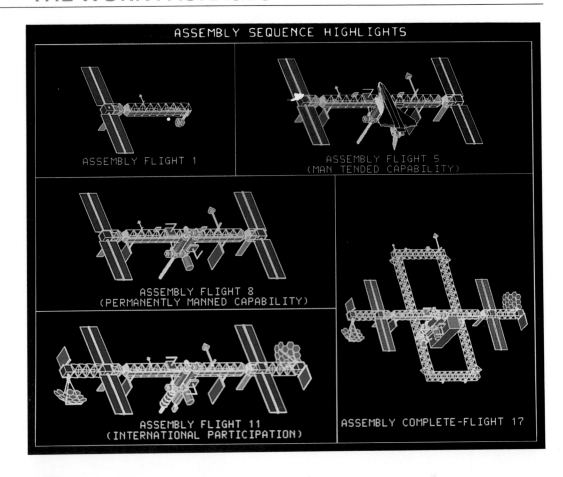

ASSEMBLY SEQUENCE HIGHLIGHTS

ASSEMBLY FLIGHT 1

ASSEMBLY FLIGHT 5
(MAN TENDED CAPABILITY)

ASSEMBLY FLIGHT 8
(PERMANENTLY MANNED CAPABILITY)

ASSEMBLY FLIGHT 11
(INTERNATIONAL PARTICIPATION)

ASSEMBLY COMPLETE-FLIGHT 17

Computer generated views of various stages of space station assembly show the configuration of the facility after the first, fifth, eighth, eleventh, and seventeenth shuttle flights.

ompany

# For Sale: Space

To arrive at the goals of the Space Station Program, NASA made a great effort to include all potential users in its planning. This included science and applications communities, the aerospace industry, and commercial organizations in the United States and abroad.

In April of 1987 NASA fired the starting gun and called for bids for the work packages from potential contractors. The contractors were asked to submit two proposals: one to cover the initial shape and design with additions gradually added on to enhance the original capabilities, and one to cover the construction of a fully developed station. Each work package is extremely varied; therefore, companies joined together as teams to compete for the contracts (see pages 50–53). The competition, which

**A**s part of President Reagan's directive to build a permanently manned space station, Lockheed Missiles and Space Company developed this configuration.

read like a Who's Who in the aerospace industry, was hot.

The competition for the costliest segment, the contract to build the Station's structural framework, estimated at $1.9 billion, was awarded to McDonnell Douglas Astronautics. The $750 million contract to construct the two United States pressurized crew modules and a laboratory was awarded to Boeing Aerospace. This contract means not only long-term commitments to a highly visible program, thus keeping the company in the public eye, but also thousands of jobs. General Electric bid to construct the external free-flying polar platform and the observational instruments anchored to it, at a cost of $800 million. Not yet awarded is the contract to provide the Station's power generation management and distribution system, estimated to cost $1 billion to build; competing are Rocketdyne and TRW.

The space shuttle carries its cargo in a payload bay 60 feet (18.3 meters) long and 15 feet (4.6 meters) in diameter. The bay is flexible enough to provide accommodation for unmanned spacecrafts in a variety of shapes as well as for fully equipped scientific labs.

Boeing Computer Services has won a minor contract, which began in December 1987, to establish the Station's Technical and Management Information System (TMIS). In 1988, NASA will award software contracts for the program's support environment and integration systems.

Several of the bidders have invested large amounts of money in quest of the desired contract. For example, Martin Marietta has stated that it has already spent at least $24 million and has built a new facility in Huntsville, Alabama, home also to the Marshall Space Flight Center, which will oversee work on the crew and laboratory modules. These two companies, Boeing and Martin Marietta, have also taken out a series of newspaper and magazine advertisements asserting that "America needs the Space Station." Boeing has been especially aggressive, perhaps due to the fact that it lost the Space Shuttle contract in the 1970s. Both companies have taken on an all or nothing attitude.

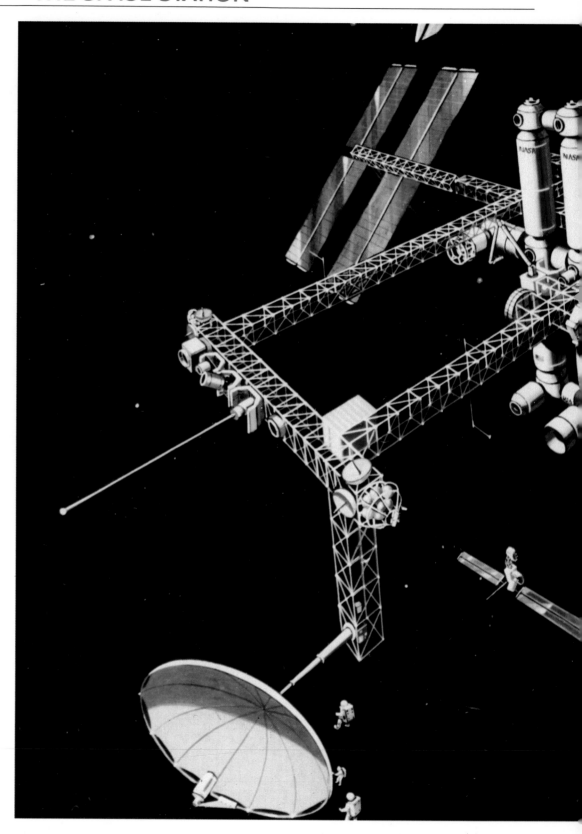

**T**his concept, furnished by NASA's Marshall Space Flight Center, is of the dual-keel configuration—a modified version of the "power tower" concept used by NASA at the start of the design program. Flying in tandem with the manned base is a co-orbiting platform.

**L KEEL SPACE STATION**

Other companies, such as McDonnell Douglas, seeking a participant's role, have submitted bids as subcontractors in other areas. Many of the bid competitors, particularly Martin Marietta, Rockwell, and McDonnell Douglas, wish to maintain a visible presence in the civilian space business because they have been front-runners in space technology from the *Apollo* Program through the Shuttle. Both Martin Marietta and McDonnell Douglas have a piece of the rocket booster business: McDonnell Douglas with its Delta rockets, Martin Marietta with the external tank and small subsystems of the Shuttle. The construction to replace the ill-fated *Challenger* is to be overseen by Rockwell.

It is widely believed that a few of the main and subcontractors will, if they lose in the major bids, incorporate themselves into the winning companies, which would ensure that the money awarded is spread among more than just one or two major companies.

Work on the Space Station is divided into four segments, each of which had two lead contractors bidding against each other for the contract. The chart below lists the associated space center, the work involved, the lead contractors, and the major subcontractors working with them.

### SEGMENT 1

Marshall Space Flight Center, Huntsville, Alabama
Work involved: crew and lab modules
Estimated cost: $750 million

| Boeing | vs. | Martin Marietta |
|---|---|---|
| Grumman Aerospace | | General Electric, Astro-Space Division |
| Lockheed Missiles and Space | | Hughes Aircraft |
| Teledyne Brown Engineering | | United Technologies (Hamilton |
| TRW | | Standard) |
| | | U.S.B.I. Booster Prod. |
| | | Wyle Laboratories |
| | | McDonnell Douglas |

### SEGMENT 2

Johnson Space Center, Houston, Texas
Work involved: framework
Estimated cost: $1.9 billion

| Rockwell | vs. | McDonnell Douglas |
|---|---|---|
| Grumman Aerospace | | Honeywell |
| Harris | | IBM |
| Intermetrics | | Lockheed Missiles and Space |
| Sperry | | RCA |
| SRI International | | |
| TRW | | |

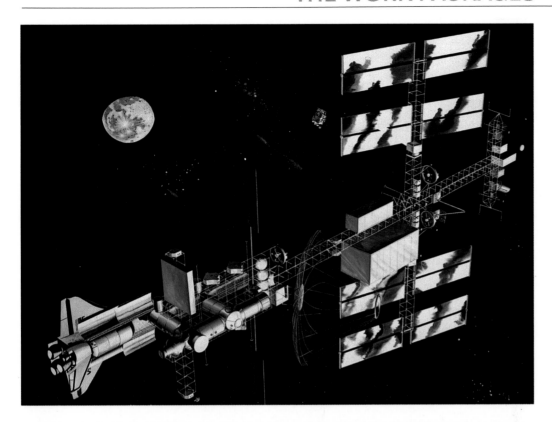

This illustration by Rockwell International shows a 50 kw power module with three other modules attached. Its antenna is aimed toward Earth.

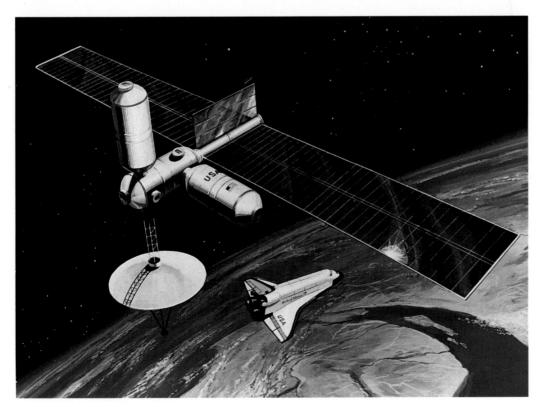

This configuration—the "power tower"—is one of a family of configurations which use similar elements or components. This concept is produced by Martin Marietta of Denver, Colorado.

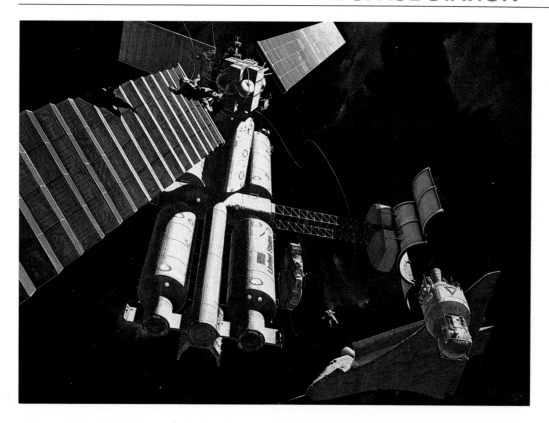

A Lockheed Missiles and Space Company space station reference design configuration. Shown here (below left) are two approaches to the space station power systems currently being studied by Rockwell International's Rocketdyne Division. Both systems would provide power for the manned station. Pictured here (below right) is one possible TRW design of the unmanned space station platform.

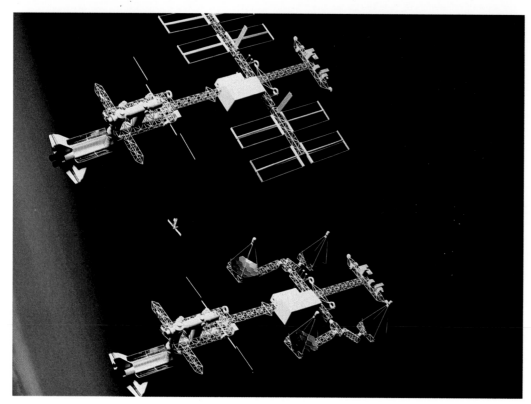

### SEGMENT 3

Goddard Space Flight Center, Greenbelt, Maryland
Work involved: free-flying platform and research equipment
Estimated cost: $800 million

| General Electric | vs. | RCA |
|---|---|---|
| TRW | | Honeywell |
| | | IBM |
| | | Lockheed Missiles and Space |
| | | McDonnell Douglas |
| | | Computer Sciences |

### SEGMENT 4

Lewis Research Center, Cleveland, Ohio
Work involved: power system
Estimated cost: $1 billion

| Rocketdyne | vs. | TRW |
|---|---|---|
| Ford Aerospace and Communications | | Lockheed Missiles and Space |
| Garrett Fluid Systems | | Planning Research Corp. |
| General Dynamics | | Analex |
| Lockheed Missiles and Space | | Teledyne Brown Engineering |
| Sundstrand | | Eagle Engineering |

# The Brick Moon: Early Designs

An 1869 *Atlantic Monthly* article, "The Brick Moon," penned by Edward E. Hale, detailed one of the earliest plans for a space station. The station proposed by Hale was round, 200 feet (61 meters) in diameter, and made of bricks. Housing a hefty crew of thirty-seven, Hale's brick moon was to orbit the Earth at 3,700 miles (1,850 kilometers) and would aid in ship navigation. In order to communicate with control on Earth, Hale theorized that the station's occupants would jump up and down on the station's exterior surface, creating signals in Morse code.

Konstantin Tsiolkovsky and Herman Oberth, two early space-flight theorists, both discussed the idea of stations in Earth orbit. As early as 1911, Tsiolkovsky considered the concept, and in 1923 the Russian wrote of orbiting "at a distance of 2000 to 3000 versts (a Russian unit of distance equal to 0.66 miles [1.22 kilometers]) from the Earth as its Moon." Oberth, in his book, *The Rocket*

*Into Interplanetary Space,* was not only the first to describe an orbiting manned satellite as a "space station" but also believed that it might be utilized as an Earth observation site, world communications link, weather station, and as an orbital refueling stop for space vehicles bound elsewhere.

During the 1950s, Werner Von Braun conceptualized a large space station. His ideas were, in fact, widely touted in the March 1952 issue of *Collier's* magazine. In a regular feature of the publication entitled "Man Will Conquer Space Soon," Von Braun wrote a piece called "Crossing the Last Frontier" in which he proposed a triple-decked, 250-foot (80 meters) wide, wheel-shaped station located in polar orbit. This space station would be a "superb observation post" from which "a trip to the Moon itself will be just a step." Von Braun's primary station would be accompanied by a free-flying, unmanned but "man-tended" astronomical observatory, meaning that the station would have company in the form of "space taxis" or "shuttle crafts" that would dart about to ferry spacemen and equipment from resupply ships to the space station.

**A**n artist's conception of a possible station configuration based upon studies conducted by TRW.

Other early station designs envisioned include:

In the 1920s, Australian Herman Noordung developed a 100-foot (30-meter) "living wheel" stationed in geostationary orbit.

A Manned Orbiting Research Lab (MORL) proposed by Douglas Aircraft in 1964, using a two-man Gemini spacecraft to house an orbiting station. The idea was dropped in lieu of other projects.

The United States Air Force's Manned Orbiting Lab (MOL) which was ready to conduct military operations in space until new technologies and budget cuts in defense spending contributed to its demise in the late 1960s.

NASA's *Skylab* space station, which utilized leftover Apollo hardware. It was discontinued due to an accident and to budget restraints.

NASA's "Phase B" modular space station concept, developed in the early 1970s; this was the first proposed station designed to be assembled from modules launched into space with the Space Shuttle. This plan was taken off the boards due to the emphasis placed on the Shuttle system and changing technology.

NASA's Manned Orbital Systems Concept (MOSC) space station, developed in 1975, utilizing *Spacelab*-derived elements to minimize the overall cost of the station.

NASA's Science and Applications Space Platform (SASP) station concept, developed late in the 1970s. This plan made use of existing technology to develop a small, expandable station which would be utilized to conduct science applications and technology missions. By this time, however, NASA was seeking a more far-ranging program.

SPACE PLATFORM
POWER MODULE, MDA, ET

This artist's rendering shows the space shuttle's external tank utilized in Earth orbit as the primary use of a space platform.

Over the past twenty-nine years, numerous designs and ideas have been studied and proposed by NASA. These designs have ranged from housing two to twelve crew members in space, with crew-duration periods from a few weeks to years. These concepts were fashioned by two considerations: the launch vehicles available to boost the station into orbit, and its intended use. Early designs were affected by the uncertainty of human performance in space, particularly due to the factor of zero-gravity. Therefore, today's emphasis in station studies evolved from general re-search and, most recently, from commercial operations in space.

In the 1960s and 1970s, the station ideas were centered around experimental and scientific data collection. Some early objectives were actualized with unmanned satellites like *Telestar,* the *Voyager* probes, and the *Spacelab* flights aboard the Space Shuttle. After the late 1970s, space station thinking included such new tasks as assembly and manufacturing. Space station mission planning continues to emphasize scientific exploration; however, the emphasis is on making the Station more cost-effective.

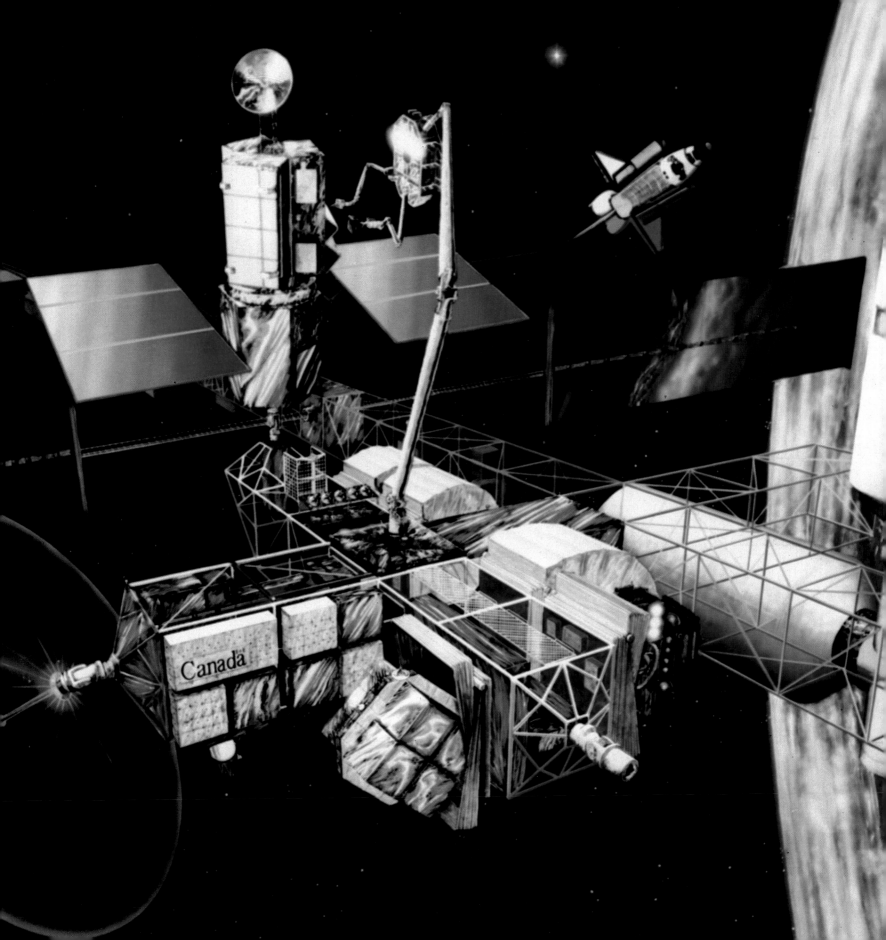

# International Participation

President Reagan's January 1984 State of the Union address extended an invitation to friends and allies of the United States to participate in the development and use of the Space Station (see page 23). NASA's long-standing partners have participated in the planning and development, and have expressed interest in the operation and use of the Station. The European Space Agency,

Canada, and Japan have conducted, at their own expense, studies parallel to NASA's. The United Kingdom is also examining what it could potentially contribute to the program.

In 1984 NASA held a series of workshops attended by all potential international partners, to plan space station activities and to exchange information. Here, negotiations were initiated with the European Space Agency, Canada, and Japan for cooperative ventures. Many problems, such as how to use the Station equitably, must still be hammered out at the negotiation table between potential partners. As a framework, NASA has established three criteria for its partners in the Space Station. They must all be builders, users, and operators. NASA has stated that it doesn't wish to repeat the *Spacelab* situation in which the ESA built the laboratory module for the Shuttle but did not operate it.

A key issue is who controls the use of the Station's modules. NASA's earliest proposals have been discarded by all partners. The United States position that the Station be managed by a multilateral board, chaired by NASA, that would plan and coordinate operations, with NASA making all decisions on issues where the cooperating agencies are unable to reach a consensus, has been strongly disagreed with. The other agencies do not wish to relinquish final authority for control of their own Station elements, even though the ESA believes that NASA should have control in emergencies. The ESA wants control over the *Columbus* pressurized module (see page 69) as well as the two European platforms.

The current proposal permits each partner to make use of its own Station infrastructure in accordance with the op-

**P**revious page: this concept is one of those proposed in the development of an Integrated Servicing and Test Facility as Canada's contribution to the Space Station. The facility would include a new generation of remote manipulators with advanced sensors, a robotic servicer work station, and special tools.

**L**eft: A model of the space station reference configuration. The astronaut at the base provides a sense of scale. Right: A 70mm camera, aimed through the flight deck's aft window, onboard the Earth-oribiting shuttle *Challenger*, recorded this clear view of the Canadian-built Remote Manipulator System (RMS) arm.

erating agency's interpretation of peaceful purpose; such use may include national security. NASA would, however, have overall responsibility for managing the Station's manned core.

Officials from all partner agencies have discussed plans for usage, but are unwilling to project how much Station operations and use will cost, other than to state that the project is extremely expensive. Despite a lack of hard information, the partners plan to sign, before 1988, cooperative agreements that will commit them to pay, maintain, and operate their own hardware elements. The Memorandum of Understanding (MOU) which each foreign space agency will sign with NASA to govern international cooperation, contains a formula for dividing up such resources as generated

Astronauts Bruce McCandless and Joseph P. Kerwin in the neutralized gravity environment of the Neutral Buoyance Simulator. This tank was employed by the astronauts to evaluate the methods used to service large beams in space.

electrical power, communications, and user accommodations. NASA proposed that after housekeeping resources are brushed aside, the remaining resources be divided as follows:

NASA-attached payloads would account for twenty percent of the resources set aside.

Canada would be awarded thirty percent of the resources.

The remaining payloads would be divided equally among the United States, Japanese, and European laboratories.

Within the labs, the United States would get one hundred percent of the NASA lab resources and fifty percent of resources in both the Japanese and European labs.

Space Station crew allocation will utilize the same formula as would the division of common operations costs between the partners.

# The Canadian Factor

Canada's space resources are small compared to those of the United States, although the Canadians, like the Japanese and Europeans, appear to be more geared toward early use of the Space Station than the United States. Since 1984, Canada has spent $5 million (Canadian) and submitted twelve industry contracts for Space Station-user development work. According to the Canadian National Resource Council, the government has earmarked $50 million specifically intended for the development of the Space Station over the next several years. Concentrating on use that will bring an early return, Canada will, in the next several years, augment

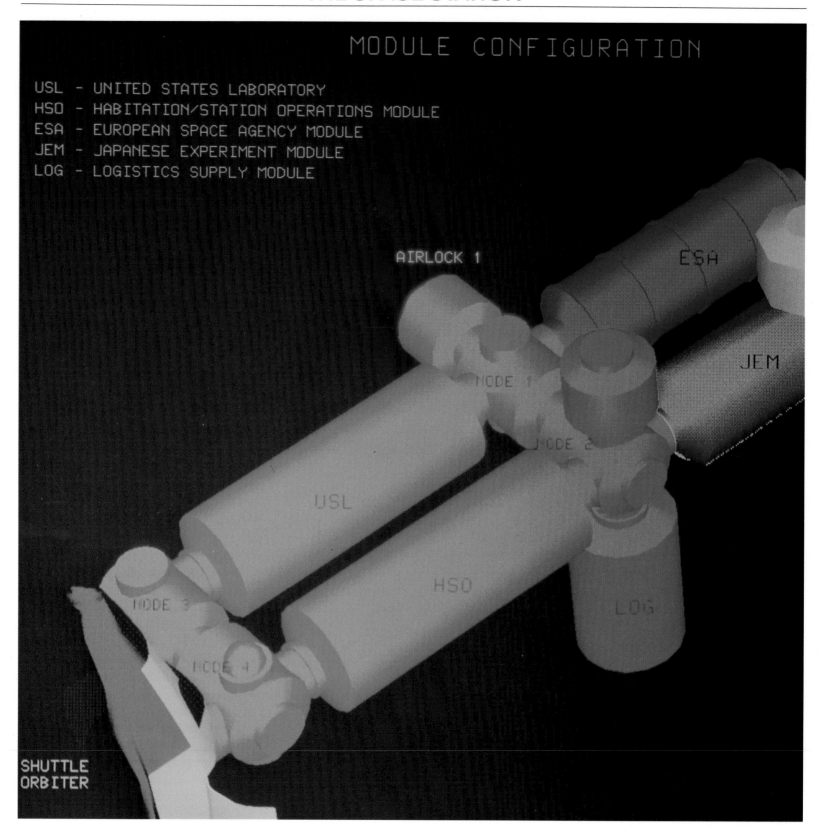

MODULE CONFIGURATION

USL – UNITED STATES LABORATORY
HSO – HABITATION/STATION OPERATIONS MODULE
ESA – EUROPEAN SPACE AGENCY MODULE
JEM – JAPANESE EXPERIMENT MODULE
LOG – LOGISTICS SUPPLY MODULE

A computer generated view (right) of a revised baseline configuration. Changes include larger nodes and a relocated service bay.

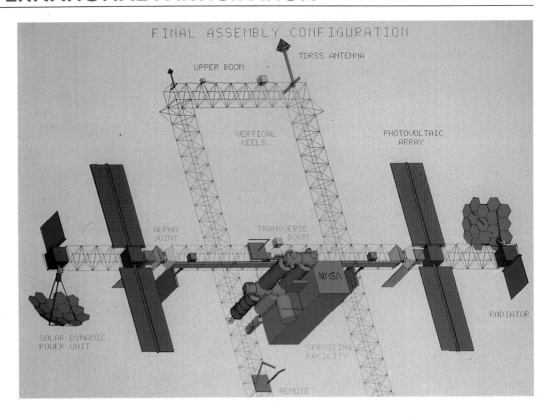

FINAL ASSEMBLY CONFIGURATION

UPPER BOOM

TDRSS ANTENNA

VERTICAL KEELS

PHOTOVOLTAIC ARRAY

ALPHA JOINT

TRANSVERSE BOOM

NASA

RADIATOR

SOLAR DYNAMIC POWER UNIT

SERVICING FACILITY

REMOTE

A computer generated rendering (left) of the module layout illustrates the modified "nodes" which connect the lab and habitation modules.

its overall space budget from $335 million to $824 million.

The motivation for this surge is the Canadian Mobile Service Center (MSC), which will be the main system used to dock to the Space Station, making the MSC the most important foreign component of the NASA program. A huge robotic manipulator arm that will be larger and stronger, and will have one more joint than the arm Canada built for the Shuttle, is being designed. The Station arm will be built to handle 200,000 pounds (90,720 kilograms), compared with the 65,000 pound (29,484 kilogram) capability of the Shuttle arm. According to the present design, the Shuttle would fly within 55 to 60 feet (16 to 18 meters) of the Station and then be soft-docked by the MSC arm, which would grasp a fixture on the Shuttle exterior and pull the two together.

The secondary docking system would involve a direct approach by the Shuttle to the Station. The problems with this plan include possible contamination of the Station by the Shuttle's Reaction Control System, as well as acceleration and docking approach speed. The MSC would be responsible for docking payloads from expendable crafts. Unlike the Soviet's *MIR* station, NASA does not plan to utilize an automated rendezvous and docking station system, which requires Earth-based monitoring.

The assembly of the Station is also dependent upon the MSC. The Shuttle's Canadarm could be utilized for the first couple of assembly-oriented missions; however, once the Station mass reaches 65,000 pounds (29,484 kilograms), it will exceed the load limit of the Canadarm, and the use of the MSC will become crucial.

Because of the primary role of the

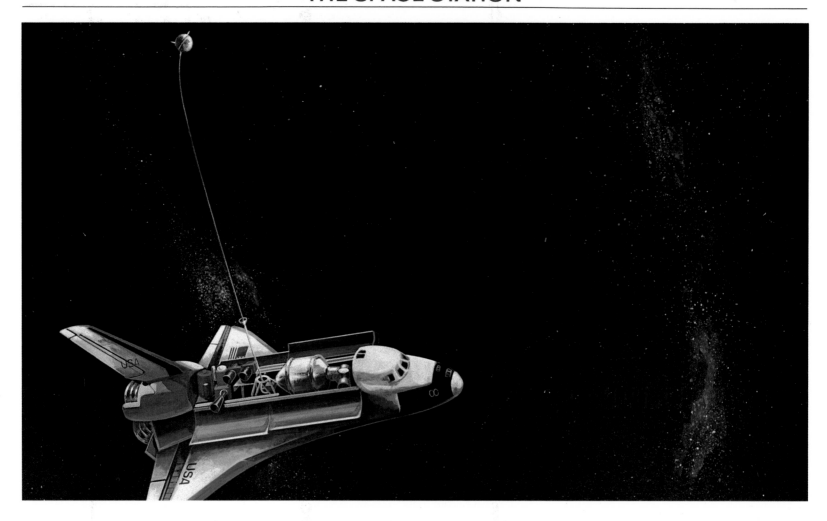

MSC, work on this system must remain parallel to the work of United States contractors. This means that if problems were to arise with the MSC, the entire Station schedule would have to be amended. By comparison, the development of the European and Japanese modules could be slowed without affecting the Station schedule.

Once NASA adopted its two-stage space program for the Station, deployment of the MSC was separated into two phases. At present, the Canadian Space Agency (CSA) and NASA Station engineers are attempting to plan the strength capability that must be constructed into the MSC during its pri-

mary phase in order to guarantee that it will be capable of lifting pressurized Station modules from the Shuttle's cargo bay and also of carrying them to their respective positions on the platform's transverse boom.

CSA and NASA believe that, in time, the MSC will be able to perform three types of motion: movement up and down the boom structure, rotation in place, and plane changes back and forth between horizontal and vertical booms. It is possible that the plane change capability may be deferred if the MSC can be used to move the pressurized modules into place. At present, NASA plans to first launch the MSC elements impor-

tant to assembly operations. This includes the arm, part of the maintenance depot, and certain control systems. The complete MSC includes the United States supplied mobile transporter and the Canadian supplied interior and exterior control stations, one or two manipulator arms, a special-purpose dextrous manipulator for servicing tasks, and a maintenance depot that will house tools and unused attachments.

For the first several flights, the MSC will be controlled from the Shuttle orbiter; later, however, the controls will be placed in a cupola. The cupola is a "hat" on one of the Station's resource nodes, which connect the pressurized

**T**his drawing (far left) shows a satellite on a tether—a super-strong cord which can be as long as 60 miles (97 km). This type of satellite would perhaps be a joint venture between NASA and Italian engineers.

**A**n astronaut equipped with an extravehicular mobility unit (left) is shown in the process of constructing and aligning a large structure in space.

lab and habitat modules. The cupola is designed to house an observation area where astronauts can conduct interior vehicular activities (IVA's) utilizing remote controls to do MSC work outside the Space Station.

Also being developed by NASA and CSA for the MSC are "end effectors" which function like robotic hands. The Telerobotic Flight System (TFS) built by NASA, will probably be the first used. The Canadian Special Purpose Dextrous Manipulator (SPDM) will be deployed several years later, as the development of the TFS and the SPDM will most likely result in tremendous advances in automation and robotics technology.

All of these highly technical chores will need men and women to oversee them. The CSA hopes to work out an agreement with NASA for one ninety-day tour of duty per year for a Canadian astronaut. At present, Canada has a six-member astronaut corps that was originally selected for Shuttle flights. In the future, an open competition will be staged in Canada and those chosen will be assigned to assist the MSC. Since a major emphasis of the MSC is on extravehicular activity (EVA), along with life sciences and materials processing, Canada expects NASA to permit Canadian astronauts to perform EVA on the Station.

# The European Factor

Unlike that of Canada, European participation in the Station is viewed as a stepping stone to the eventual establishment of an autonomous European Station around the year 2000. The European Space Agency (ESA) has two separate station designs, one autonomous, the other NASA-linked. Both of these designs are called *Columbus*.

The two parts to be built by the ESA for the NASA Station will consist of a manned module, to be launched with the first several flights of the Shuttle, and later, a serviceable unmanned platform. The *Columbus* proposal is a joint effort by West Germany and Italy. Central to the project is the use of the *Spacelab* orbital module developed for the ESA by the major West German aerospace firms, Messerschmitt-Boelkow-Blohm (MBB) and ERNO, along with Italy's Aeritalia, a space design organization.

Studies conducted by MBB, ERNO, and Aeritalia conclude that *Columbus*, a laboratory, could evolve into several different configurations by using pressurized *Spacelab* modules as its basic building blocks, assembled and incorporated with other modules, systems, and equipment.

Through the Columbus Utilization Preparation Program, the ESA is currently teaching Europeans how to operate Station hardware. At present *Columbus'* principle plans and missions include a manned laboratory docked to, and serviced by, the Station; and a manned free-flying platform, possibly to be built in orbit by British Aerospace, and capable of being attended to from the Station or Shuttle by an autonomous

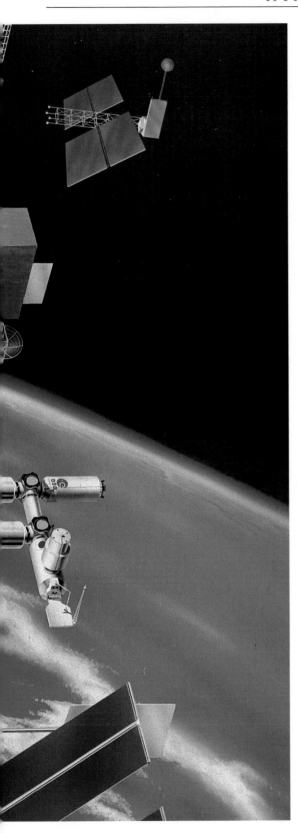

An artist's interpretation of the "dual-keel" station concept. Also shown are the proposed modules furnished by the European Space Agency, and the proposed Japanese Experiment module.

propulsion system. It will be equipped to house European work crews; however, ESA has yet to determine what *Columbus'* exact use will be.

# The Japanese Factor

In hopes of raising its profile in the international science community, Japan is extremely eager to participate in the Space Station program. Since 1983 Japan has held a half-dozen station mission requirement workshops, sponsored by the Japanese National Space Development Agency (NASDA). NASDA has received 327 Station-user proposals in six major areas: science observation (5 percent), communications (5 percent), Earth observations (13 percent), technology development (26 percent), life sciences (28 percent), and microgravity (24 percent). Thirty percent of the proposed missions were from Japanese industry.

The Japanese proposal to NASA is in two parts. The experiments module would harbor facilities for materials processing, as well as for the space sciences, life sciences, and other research. The 33 by 13 foot (10 by 4 meter) module would be launched by the United States Space Shuttle. Following that, the Shuttle or a Japanese H-2 booster would launch the 26 by 6 foot (8 by 2 meter) exposure platform into orbit, which would then dock to the outside of the module. The Japanese could then exchange various sensors on the exterior platform using the robotic manipulator arm. The inside of the Japanese module would resemble the European *Spacelab* module.

# SPACE SOLAR POWER STATION
## THERMAL

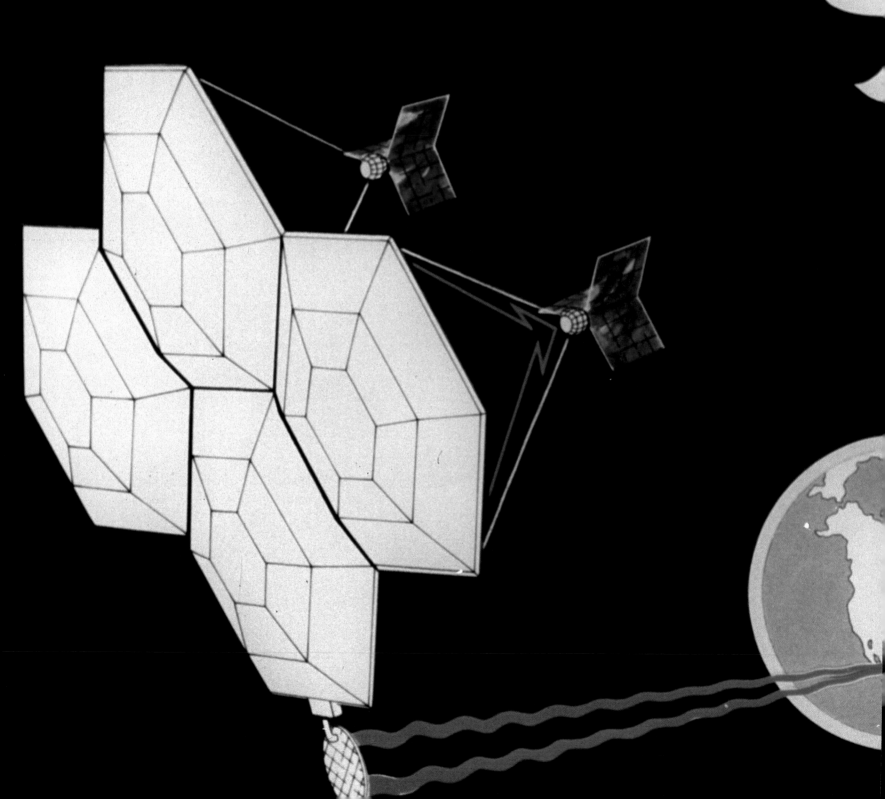

# CHAPTER 5

# Advanced Technologies

Advances in space technology are imperative as NASA prepares to assemble the Space Station, especially if it is to be an evolutionary system. Progress is essential for the conception and design of scientific missions, as well as for sustaining human life on a long-term basis.

NASA is developing a technology that is both efficient and

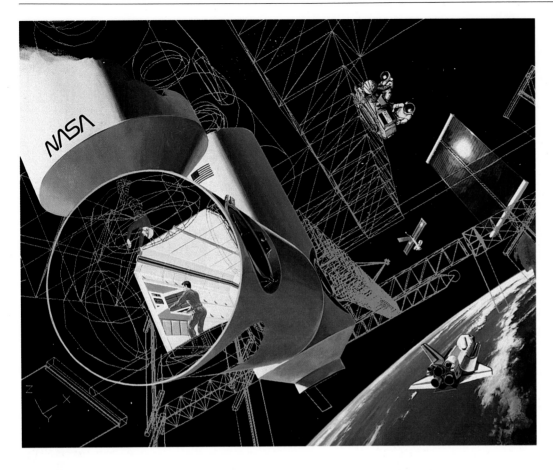

**P**revious page: A Thermal Space Solar Power Station which would generate about 14,000 Megawatts of power.

**T**he interior of a space station module (left) is depicted in this design by McDonnell Douglas Corporation. Outside the module, astronauts perform a space walk. Right: This design of the Advanced Space Operations Center depicts the shuttle unloading some of the modules which would comprise the system.

productive in its use of the Space Station. Looking toward the evolution of the manned and unmanned Station and platforms of the future, NASA has targeted areas with the highest priority and the greatest potential to increase Station productivity. These areas consist of data management, environmental control (life support), thermal control, power, and automation and robotics.

# Data Management

Data management involves newly evolved and advanced systems for data collection, storage, and retrieval. These will include data processing networks

and data storage devices. NASA is attempting to integrate high-speed circuit technology with data processing systems in order not only to guarantee that all systems are compatible with promised advances in space systems, but also to allow the continued evolution of the space platform systems as time and technologies evolve.

The IBM Federal Systems Division of McDonnell Douglas is responsible for the Data Management System (DMS). This system will integrate computer hardware, software, crew work stations, mass-storage, and networks to provide a growth-oriented and malleable information system to aid the needs of the astronauts. This multipurpose system will also provide the ability to process, acquire, store, and display information for

the major systems and payloads, as well as serve as a communications network to permit all the onboard elements to "talk together" as an operational entity. Despite the fact the IBM computer systems have been a part of every United States manned spaceflight program from Project Mercury in 1959 to the present-day Space Shuttle, the DMS is the first large-scale distributed processing network for continuous use in space.

Lockheed Missiles and Space Company was awarded the $141 million contract to develop computer software for the Space Station. This six-year contract, called the Software Support Environment (SSE), continues into 1993 when Space Station assembly in the 200-mile (320 kilometer) Earth orbit is scheduled to begin. Lockheed's job is to

This drawing, typical of the studies conducted by Boeing Aerospace Company, shows the inside of one of the two living modules envisioned for the center.

design the Station's software specifications, standards, procedures, policies, and training materials, which are to become the "tools and rules" for the various computer programs to be used in orbit and on Earth. This work will occur at the Lockheed facility, near the Johnson Space Center, and other NASA centers, with the overall management tasks to be performed by NASA's Space Station Program Office in Reston, Virginia. Most of the work will be done in Houston. Lockheed, however, also plans to seek support from its laboratories in California. At the beginning of the program, an interim software system will be utilized by NASA and the work package contractors until the permanent system is installed.

# Life Support

While details at this time are vague and will remain so until the work packages are awarded, an Environmental Control and Life Support System (ECLSS) will provide the astronauts with a breathable atmospheric mix and supply water for drinking, bathing, and food preparation. The system will also remove contaminants from the air and process biological wastes. A "closed" system, the ECLSS

will allow oxygen to be recovered from the carbon dioxide exhaled by the astronauts. Supply water, urine, and the water formed from condensation will also be reused. Food and nitrogen will be the only things that will have to be periodically resupplied to the Space Station.

The four pressurized modules of the Space Station crew base will be linked together by resource nodes. These nodes are larger, outfitted versions of the tunnels and nodes that served solely as passageways in the original baseline configuration (see page 28). The new design will add about 8,400 cubic feet (238 cubic meters) of usable pressurized volume to the manned base. These resource nodes will improve crew safety and, it is believed, productivity. With the nodes, the Space Station habitation base will have approximately 31,000 cubic feet (890 cubic meters) of usable pressurized volume.

Also being developed are higher productivity elements and superior extravehicular activity (EVA) systems which will allow for greater comfort. Higher pressure space suits will eliminate the astronaut's need for prebreathing. Other elements include a regenerable EVA mobility unit and backpack maintained in orbit, an advanced manned maneuvering unit, and a crew capsule for crew access to higher orbits.

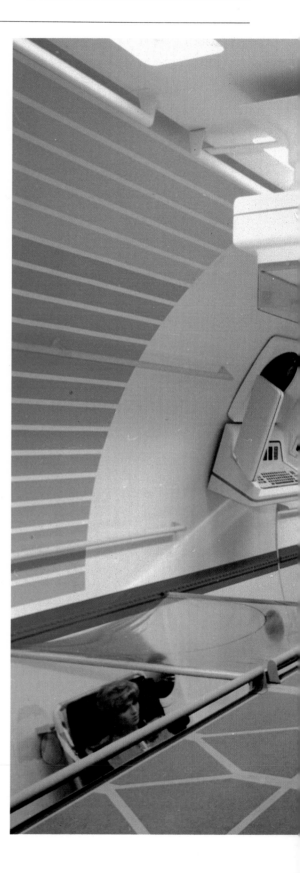

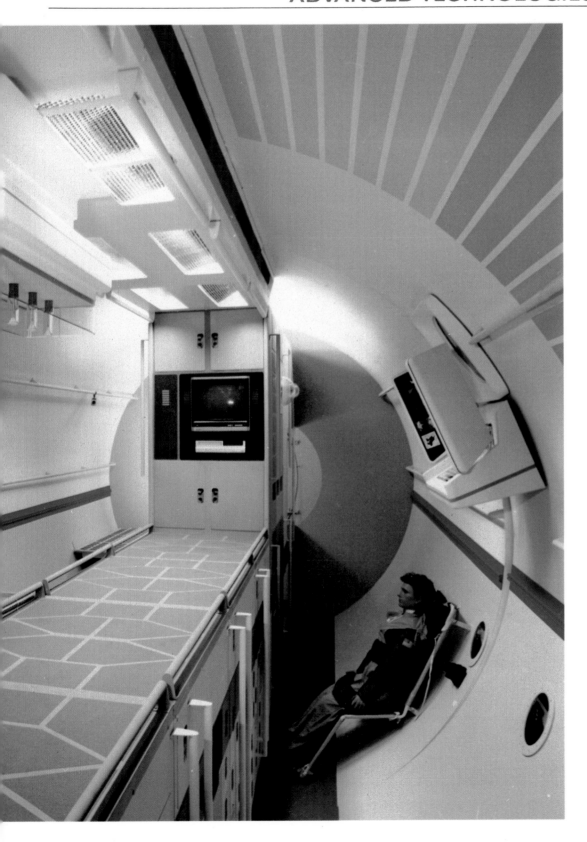

"Life in Space" could be the name of this Rockwell International mock-up of the inside of a module that provides the pressurized environment for the flight crew members.

This sequence illustrates how the orbiter would provide a necessary role in assembling antenna modules and deploying a receiving antenna for a 500 kw solar array structure.

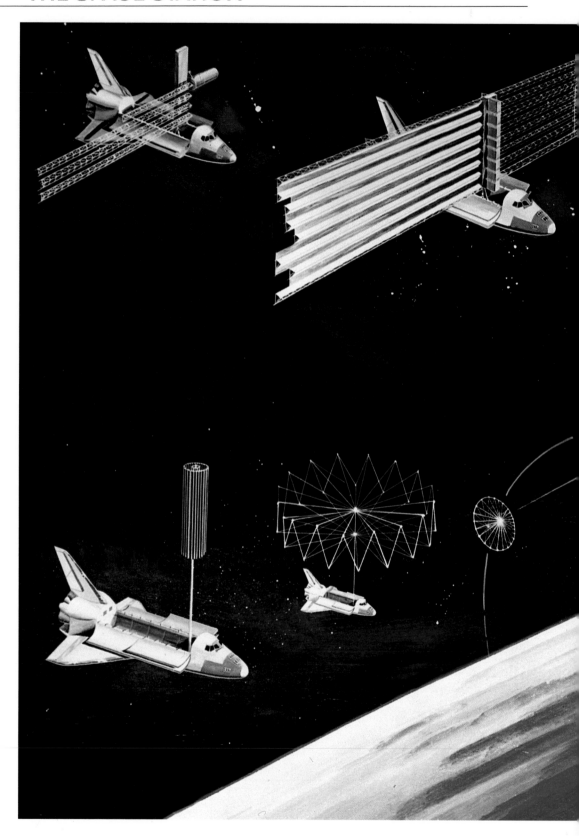

# Power

In the past, the maximum electrical power available in a spacecraft has been tens of kilowatts, with inefficient power to weight ratios rated as low as five watts per kilogram (a kilogram equals approximately 2.2 pounds). Costs have been as expensive as $1,500 per watt. NASA has been conducting extensive research on photovoltaic cells and arrays (see page 43) and they firmly believe that scientists will soon provide substantially greater and more efficient power levels and lower system costs and weights. With the development of arrays for solar electric propulsion systems, NASA expects to demonstrate a power-to-weight ration of 66 watts per kilogram. NASA believes further development will provide array outputs of 100 or more kilowatts, with a power-to-weight ratio for things like high-performance silicon arrays of 300 watts per kilogram, and with costs lowered by a factor of five or more ($300 per watt). Presently the agency and its contractors are studying ways of improving photovoltaic materials and better designs to provide cells and arrays with a higher concentration efficiency, structures of lighter weights, and the important ability to maintain high performance for extended periods in space.

NASA believes that solar dynamic power systems offer advantages over other solar conversion systems, particularly for power levels greater than 300 kilowatts. This is because of their substantially smaller areas and, therefore, a lesser need for station-keeping propulsion. This type of system will probably be installed in the initial Space Station.

Because high-power systems generate substantial amounts of waste heat

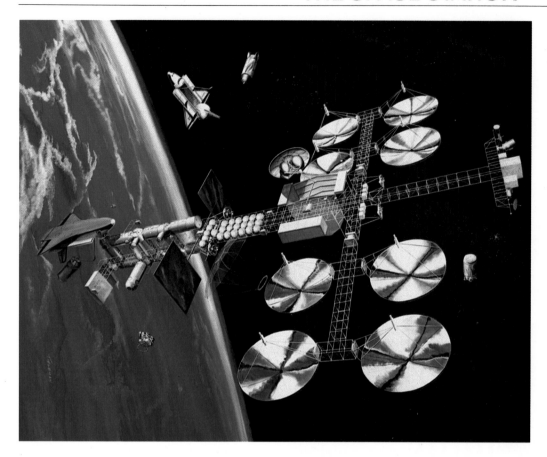

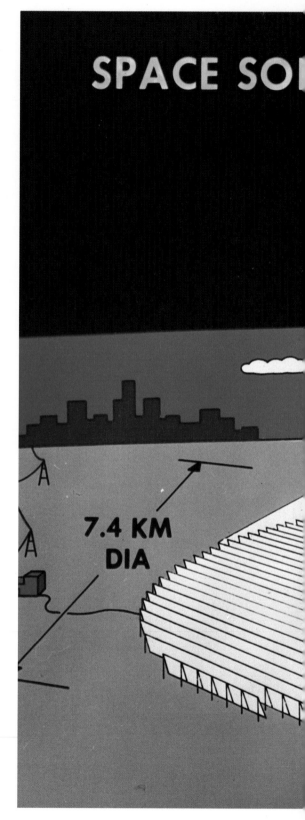

that must be disposed of and radiated into space, they require efficient thermal management systems which possess high capabilities and extremely low vulnerability to meteoroid-type damage. Therefore, inside the Station, cold plates will be integrated into thermal, power, and handling subsystems to help cool these systems. Removal of waste energy by heat pipes and pumping two-phase fluid systems is also being considered. Liquid droplets and liquid metal belt radiators are also being studied for transferring radiating heat to space.

The Tri-Agency Space Nuclear Reactor Power System Technology Program was created in 1983 to advance the technologies deemed suitable for a space power system capable of producing 100 kilowatts or more power, while

operating at full power for at least seven years. Three concepts are under investigation at this time: an in-core thermionic system, a lithium-cooler reactor (a fairly simple thermo-electrical energy conversion system), and a low temperature reactor, which would utilize existing liquid-metal, fast-reactor technology enhanced with a dynamic (meaning equipped with free-moving pistons) energy conversion system which would convert the liquid-metal into energy. Before a prototype unit is assembled, a single system concept will be selected.

As the Space Station is enlarged, it will require power management and distribution at high voltages and probably at high alternating current frequencies. Because of this, the Station will need materials resistant to higher temperatures

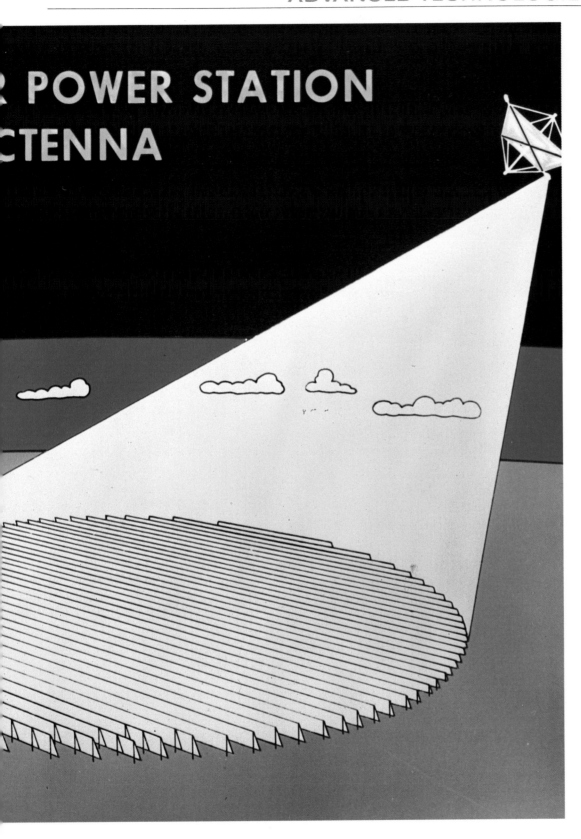

R POWER STATION
CTENNA

Far left: A station configuration using a solar dynamic power generation system, as a means of generating the station's electrical power. Left: An artist's conception of the ground-based microwave receiving station proposed by the Raytheon Company.

and radiation for circuits, electronic components, switching networks, transmission lines, high-speed switches, short-term energy storage, charge indicators for batteries, autonomous operation, and control of interactions between high voltages in space. Extremely important factors in the reliability, weight, and life of high-capacity space power systems are energy storage systems. The technology is currently being developed for systems based on nickel and hydrogen electrochemistry, as well as for regenerative electrolyzer systems for fuel cells. NASA's objective is to design reliable orbital systems capable of storing hundreds of high voltage kilowatts of energy. A regenerative fuel cell, compared to a battery system, is lighter and considerably more flexible. Alkaline and acid technologies are being studied for their performance and endurance characteristics in connection with both fuel cell and the electrolyzer elements of fuel cell systems.

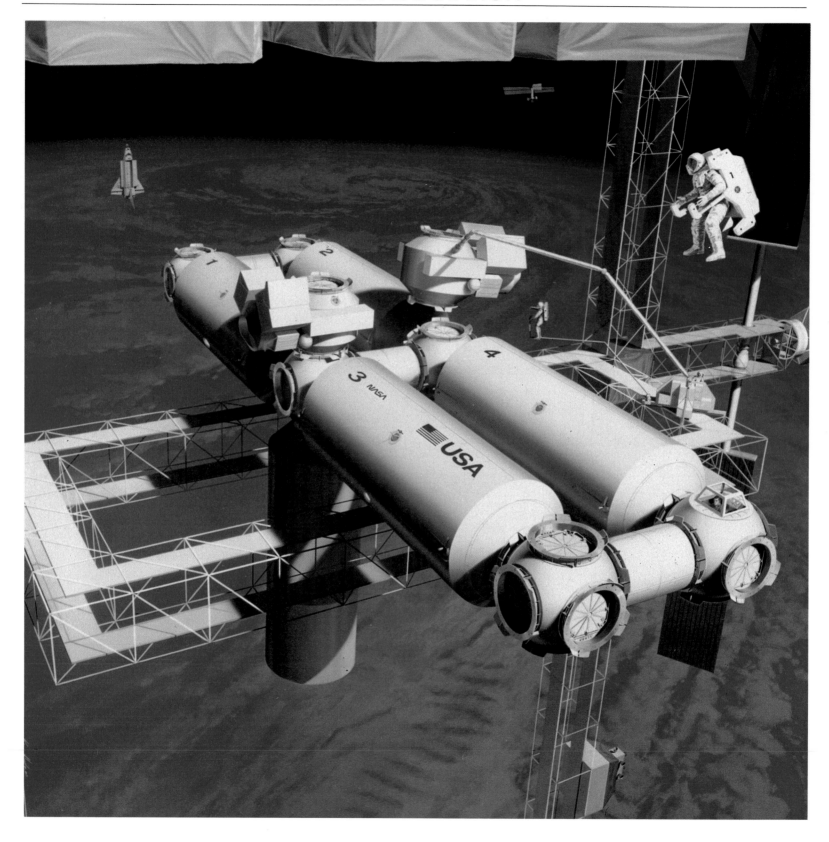

**R**ight: A Martin Marietta design for the completed space station.

**L**eft: A McDonnell Douglas rendering of the station shows a robotic arm being controlled by an astronaut. The arm is moving a new part, called a connective node, into place.

# Thermal Protection

Several types of materials are being studied to protect vehicles against temperatures exceeding 2,300 degrees Fahrenheit (1,210 degrees Celsius). Ablating materials (those that burn away with heat), unable to be reused, are of least importance. NASA is studying advanced carbon-carbon composites which can protect against temperatures up to 3,000 degrees Fahrenheit (1,649 degrees Celsius) and could be used in advanced vehicles, especially in the nose region and on their control surfaces. Another alternative for protection against higher temperatures is the use of advanced ceramics, similar to those used on the Shuttle orbiter, but with greater durability and higher temperature resistance than those currently used.

This Rockwell International drawing depicts the manned phase of the Station with the addition of the solar dynamics power system.

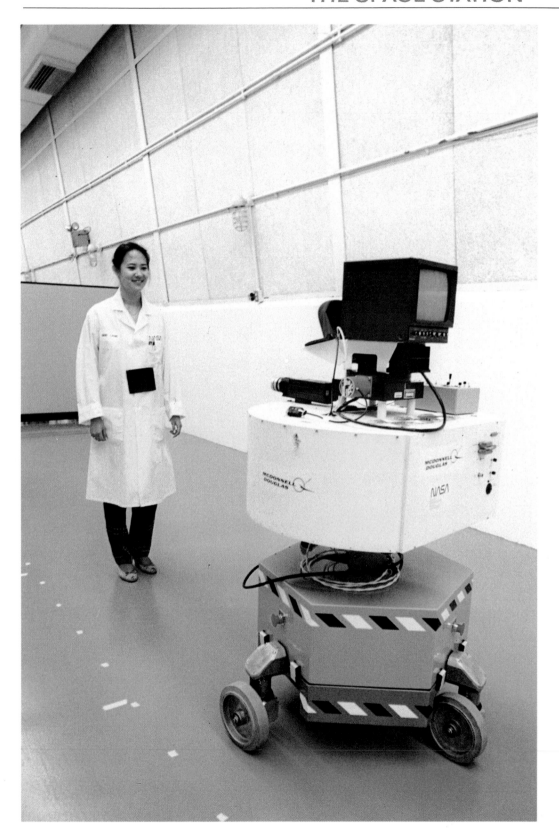

**A** robot uses a McDonnell Douglas-developed sensor on top of a JSC developed base to follow Ann Marie Ching. The robot uses a video sensor to follow Ms. Ching by tracking the contrasting black square on her white lab coat.

# Automation and Robotics

This program's objective is to provide technology to augment, through automation, the capabilities of systems and human productivity in space. Because NASA's missions are becoming increasingly complex, the requirements of conventional technology are making spacecraft approaches unaffordable and often, unachievable. When the mission objective is the ability to land on free-floating space platforms, this inadequacy becomes especially obvious, due to the lack of control possible on a platform.

In order to meet the long-term challenges of the Space Station, the degree of autonomy in space operations must be substantially increased by automated systems able to perform remote manip-

**R**obots like the one pictured here will be used to work on the Space Station. Tasks such as satellite repair will be completed by robots to reduce the risk to astronauts. The robots will use many "man-rated" instruments like the Maneuvering Unit, shown here on the robot's back.

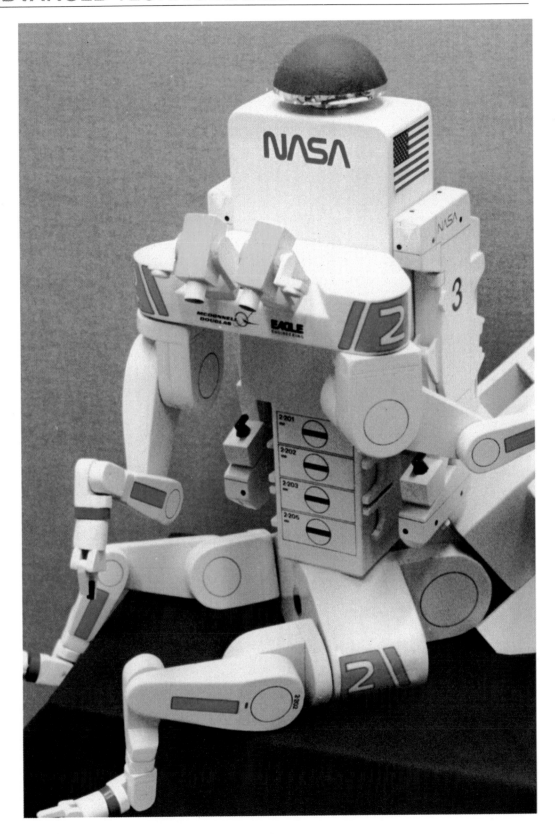

ulations and systems control in space. When space systems are automated, mission costs are reduced, scientific and engineering capabilities are assisted, and new opportunities are attainable, creating optimal utilization of both the human and machine force in space. As the Space Station is designed, contractors are particularly interested in the recommendations of the NASA Advanced Technology Advisory Committee, which is targeting automation and robotic technologies for use in the Space Station.

The major areas of technology involved in this program are sensing, planning, decision-making, and space operations using human supervised as well as autonomous systems. NASA's program will integrate basic technologies to yield a basis for understanding and evaluating the tactical situation. It will also exercise supervisory control over remote manipulation and mobility requirements. Ac-

One of the four Brayton-cycle power generating units of the Thermal Solar Power Satellite. The total satellite weighs over 50,000 tons and will cost billions of dollars, but could greatly reduce the kilowatt-hour price of energy to make it economically competitive in an energy-limited world.

tivities benefiting from automation include maintenance and operations of ground-based monitoring systems and space platforms, on-site experiments and their operations, satellite servicing, large space structure assembly, repair and maintenance of Space Station-based systems, and scientific payload remote operations.

Before the first launch of Space Station elements aboard the Shuttle, NASA will have ready a flight telerobotic system, attached to a mobile remote manipulator, which will assist in assembling and maintaining the Station. This telerobotic system will be utilized as a "smart" front end on an Orbital Maneuvering Vehicle (OMV) for remote operations and for servicing free-flying payloads.

As a result of the growing concern over competition in the international marketplace, there is a tremendous interest in the United States Congress in accelerating the dissemination of advanced automated technology to, and in, United States industry. This program of automation and robotics development would not only serve as a visible demonstration of advanced automation but would also spur the passing on of the new technology to the private sector, thus establishing the United States as the leader in automated technology.

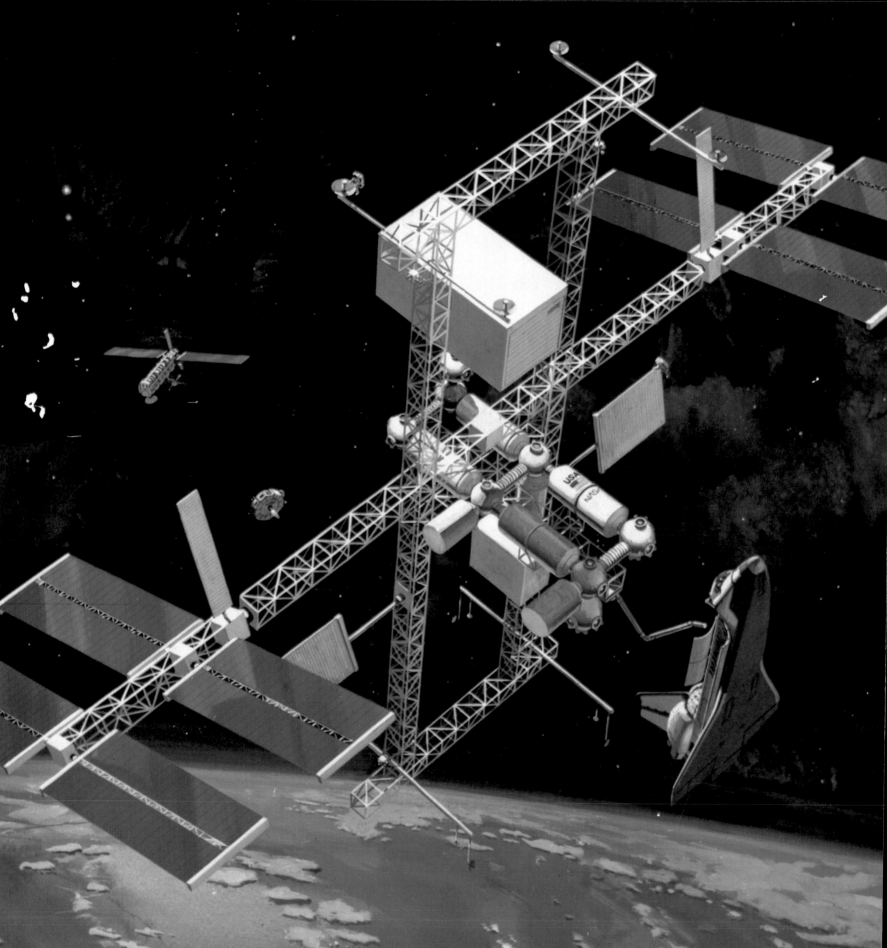

# CHAPTER 6

# Expanded Use of the Space Station

"Similarly, there should be little need to make a case for the earthly value of the space program. Space itself will eventually prove the wisdom of going there. But even for those immune to history, the space program has already proved its worth. Not just Tang and Velcro but the entire computer industry as we know it, much of today's medical technology, and uncountable lesser benefits have been developed by it."

—Hugh Downs, a prominent space advocate, comparing the Space Station to the Soviet station, *MIR*.

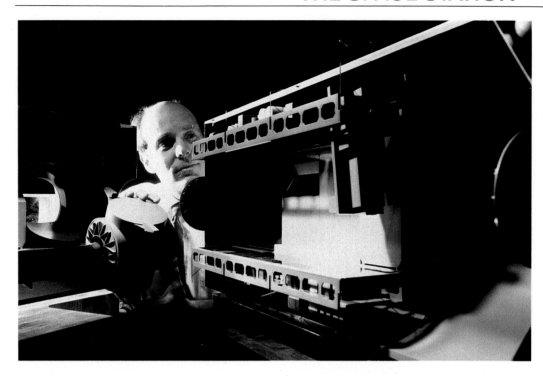

**P**revious page: A "dual-keel" configuration, modified from the "power-tower" concept, furnished by Rockwell International. Left: a designer shows one possible shape for the 1/400th scale model of a proposed station module. Built by Boeing Aerospace Company, the module has been designed as a materials processing, manufacturing, and technology lab.

Since the early 1960s, the United States has invested over $100 billion dollars in equipment, resources, facilities, and information bases in order to better understand space and its potential. Through the utilization of commercial satellites, NASA and private industry have revolutionized the methods of business. By providing access and data from global communications networks, commercial satellite systems have produced a viable new industry, technology spin-offs to other industries, thousands of jobs directly or indirectly related to the space industry, and more efficient solutions to many business needs.

Through the Office of Commercial Programs (OCP), NASA not only takes an active role in targeting potential early uses of the Space Station for commercial activities, but also attempts to encourage commercial use of the Space Station. The OCP oversees an ongoing commercial outreach program geared toward attracting potential commercial us-

ers. This outreach program asks industry to consider: developing critical technologies in space which can be transferred to Earth-based business applications; the potential for producing high-value-added materials in space; and space-related products being pursued by domestic and international competitors.

NASA, stressing that the opportunities pursued in space should be related to the business needs being developed on Earth, has begun to develop many industry-related technologies. Agribusiness, for improved effectiveness of operations through better crop information is being developed, as well as computers using new organic materials for high-speed optical computer functions. Electronics are being developed to process superconducting materials in ultra-high vacuum facilities in space, as well as utilities for monitoring and controlling electrical power networks from space. Pharmaceuticals are being developed to experiment with new drug design tech-

**A**bove: This full-scale mock-up of a Space Station Microgravity and Materials Processing Facility, has been built by the Boeing Aerospace Company to study accommodations and the equipment needed to support such a laboratory.

**L**eft: A space-suited astronaut prepares to snare an OMV using the station's mobile manipulator system.

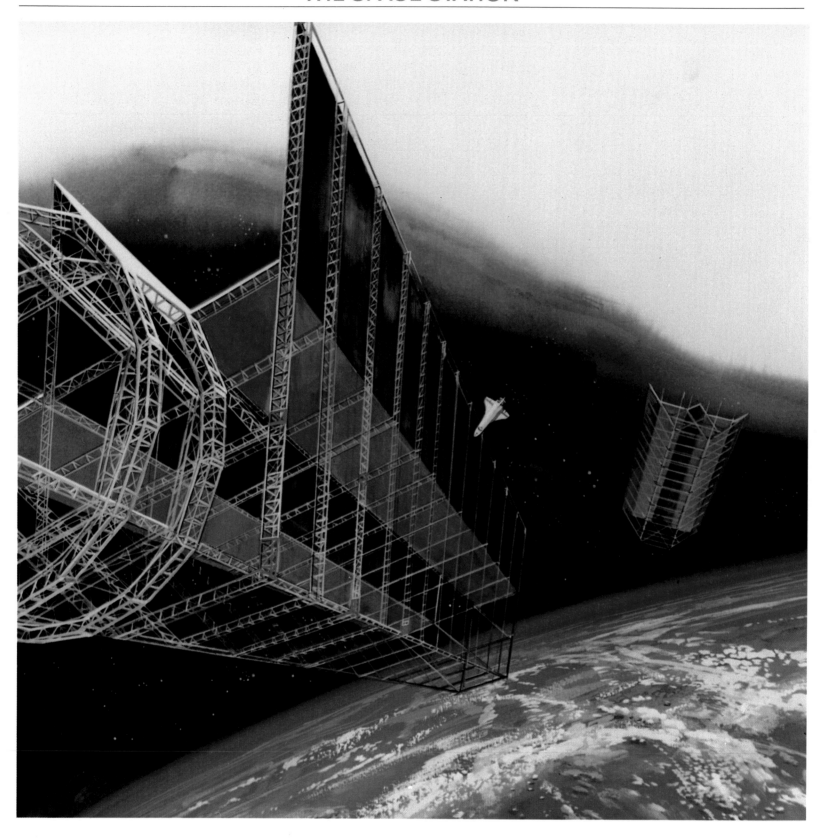

The Shuttle would deliver both the materials and the machinery required to construct large space structures, such as this demonstration satellite solar power station (left) which would beam a continuous stream of microwave energy to earth-based receivers, converting the energy to electricity.

niques resulting from the use of x-rays to study the large protein crystals produced in space. As a result, a Space Station Commercial Advocacy Group (CAG) was initiated to interact with potential commercial users of the Space Station and entrepreneurs considering the provision of space products and services to the Station.

The opportunities for financial and product profit from the Space Station appear to be quite large. Studies were conducted to assist NASA and industry in better understanding the total picture. These studies covered a spectrum of issues, from protecting proprietary commercial operations to advocating the Space Station configuration changes in order to better use its on board microgravity environment, and reviewing the use of the United States laboratory module for commercial endeavors. Enhancing performance through ad-

vanced technology is a major thrust of the Space Station design. Propulsion, structures, materials, automation, and robotics are all probable areas where Space Station technology will spill over into the private sector. NASA also is pioneering the development of an integrated capability for cost-effective production and long-life maintenance of complicated software applications. This capability, entitled the Software Support Environment (SSE), provides the tools, training, and procedures necessary to develop, verify, and evolve software for diverse uses. The SSE is responsible for industry leaps in defining and making integrated development systems operationally available to industry. Because of this the SSE is helping to change the face of economic and technological competitiveness of the United States private sector in the growth-oriented software development field.

Right: Engineers inspect the interior of a full-scale mock-up of a space station module built by Martin Marietta. This mock-up is equipped as a manufacturing technology lab.

**L**eft: This NASA drawing depicts a module and other support systems envisioned for a permanently manned station. In view is one end of NASA's "power-tower" design.

# Microgravity and Industry: A National Lab in Space

Materials processing research and production may prove to be one of the most important functions of the Space Station. Because of the extended length of orbit with an ongoing human presence afforded by the Station, this field of endeavor will be greatly enhanced. Therefore, the United States is developing a microgravity laboratory module designed to provide its users with unparalleled resources that permit various research projects to be conducted. The Space Station's microgravity environment will encourage scientists to develop new concepts in materials research and life

sciences. Scientists working on board the Space Station in pressurized modules will have the potential to develop new supercomputers by manufacturing large, flawless crystals, such as gallium arsenide (used in making computer chips). Scientists will also be able to manufacture pure biological crystals, which are required for the identification of basic molecular structure. The manufacturing potential of growing organic polymer crystals and of alloy solidification is also related to materials research.

Two functions of the microgravity laboratory will substantially effect the program's possibilities for commercialization: the use of space to gain information useful to improve Earth processes, especially chemical and biological trends of the planet (see chapter 7), and the processing of materials in space to capitalize on the weightless conditions aboard the Station. The first function

**T**his Rockwell International conceptual drawing (right) depicts a possible growth version of the "dual-keel" shape of the station. Sample payloads are attached to the upper and lower booms of the station.

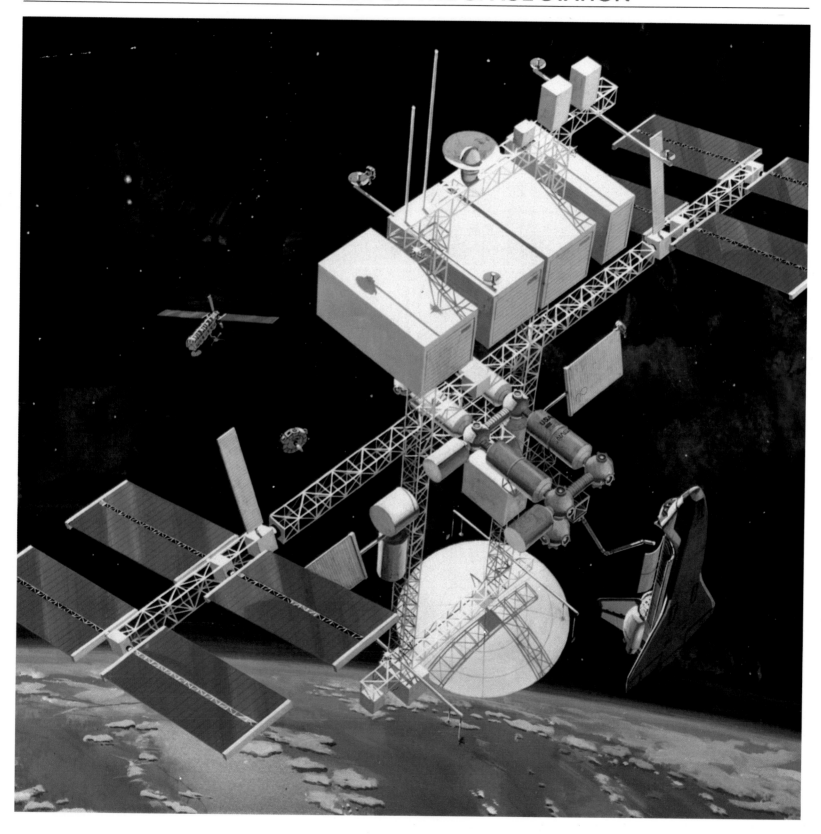

The unmanned platform, developed by McDonnell Douglas, illustrates that the rotating pallets are capable of holding both space science and applications payloads.

plans for the routine use of microgravity experimentation as a method for solving problems related to materials processing on Earth, with industry picking up most of the tab for the experiments. Several companies are interested in this type of problem-solving as the need for more accurate knowledge of the thermophysical properties of materials increases. The industrial sector is also interested in testing models of various processes by first determining whether the results from processing the material in space, in the absence of gravity, can verify that these models correctly predict the basic nature of the processes *before* integrating the complex effects of gravity. Industrial attention will be focused on growing certain crystals, such as protein, in the lab in order to better understand why they are difficult or impossible to grow on Earth. Another space activity that will be useful to many companies is the preparation of small amounts of new and unique materials yet to be developed, to either determine their characteristics or to utilize them as perfect models.

In the near future, processing samples in space will be restricted to high-value, low-volume materials such as pharmaceutical products, electronic materials, optical fibers, highly specialized alloys, and, perhaps, precision latex microspheres. As experience with the flight program increases, other new applications are expected to emerge.

This laboratory-in-space module will be launched, but only partially outfitted, on the sixth Shuttle flight, before the United States habitation module is sent into space. NASA currently intends to include nine racks of the lab module's equipment with this launch. Six of the racks are for the subsystems necessary if the module is to serve as a laboratory. Two of the launched racks are for user-

Above: A reference configuration of the station produced by the Convair Division of the General Dynamics Corporation.

A drawing of an OMV (right), a kind of "smart space tug", which would carry satellite and other orbiting objects from place to place.

support subsystems, and the remaining rack will be for user-supplied experimental equipment. The key reason why a partially outfitted module will first be launched is because of its weight. NASA has recognized that weight would be a problem and added an outfitted Shuttle flight in the assembly sequence immediately following the launch of the laboratory module. This flight would carry seventeen of the remaining thirty-five racks, most of which is user equipment, to the Space Station. Once aboard and in place, this equipment should support extended operation time for research.

The Space Station may, in time, become the commercial production site of critical materials unable to be manufactured on Earth, such as pure pharmaceuticals. Frequent crew visits will be important to assure the proper development of these production processes. Changeable payloads for commercial remote sensing instruments will also be provided for on the Station.

NASA and Space Industries, Inc. of Houston, have signed a Memorandum of Understanding. Both organizations have promised to exchange pertinent information during the second phase period. Space Industries, Inc. has planned to develop and launch via Shuttle a pressurized laboratory which would be serviced from the Space Station.

When the second stage of the Station is completed it will serve not only as a national technology and science lab for NASA and industry but also as an assembly base for new spacecraft, and as

a satellite servicing center. The expanded Space Station, featuring dual keels (two vertical spines 345 feet [105 meters] long, connected by upper and lower booms) will have two new booms, 148 feet (45 meters) in length, to carry a United States supplied servicing bay for satellites. This servicing bay will ideally be used as a permanent base for the tending, servicing, and repair of satellites and the unmanned platforms. As the technology advances, the Space Station offers the flexibility to upgrade space systems. An initial capability for placement and limited retrieval of satellites is the Remote Manipulation System, Manned Maneuvering Unit, the Extravehicular Mobility Unit (an integrated spacesuit and powered backpack), and tools for EVA. Also required will be equipment, such as holding and positioning aids, servicing, repair, and maintenance tools, berthing platforms, refueling techniques and tools, televised monitoring systems, and equipment for assembling and supporting any large structures.

The OMV is also usable, if controlled from the Space Station, to perform remote-orbit satellite services. This vehicle would probably be utilized to reboost satellites to their original operational altitudes, and deorbit satellites that have completed their periods of useful life. The Space Station will enhance in-orbit assembly and the maintenance of structures such as antennas, telescopes, lenses, and satellites before they are deployed elsewhere.

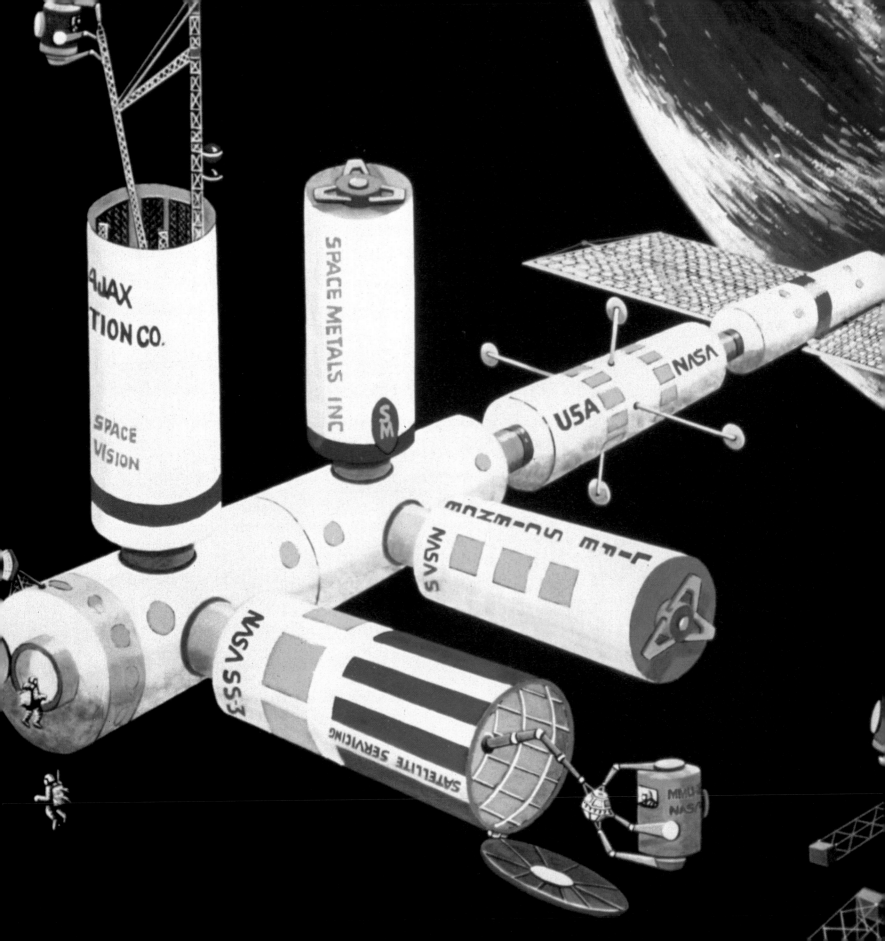

# *Science and the Space Station*

Aside from technological experimentation, the Space Station will also provide the ability to conduct space-based scientific research in such diverse fields as the life sciences, materials processing, earth science and applications, solar system exploration, astrophysics, and communications. Within the fully equipped microgravity laboratory, where commercial endeavors are tested, the Station will also

**P**revious page: An early design of a space industrial facility. Shown here are both private and public enterprises. Right: Commercial uses of space include communications, weather forecasting, navigation, mapping, and surveying Earth's resources.

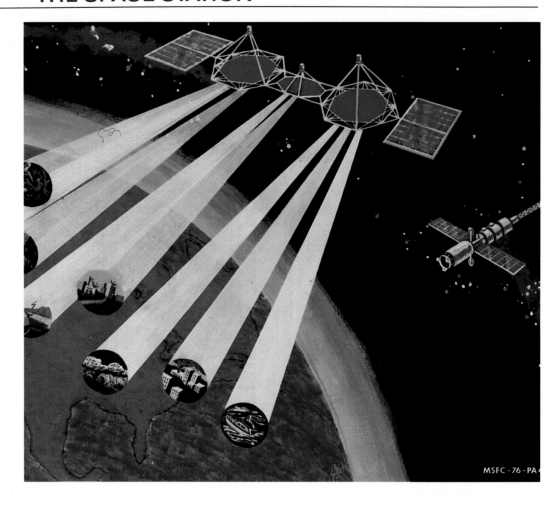

MSFC - 76 - PA

support pure scientific research. The Office of Space Science and Applications (OSSA) established a separate budget category, Space Station Integration, to maximize the utilization of the Station. This funding ($15.5 million in 1987 and $20 million in 1988) provides for technical studies, operations planning, and test laboratories. Science payloads are to be selected for the Station from those developed in the various science discipline programs. The OSSA earmarked an additional $7.5 million in 1987 for the development of science experiments to be conducted on board the Space Station. Planning for the accommodation and support of payloads is jointly coordinated between the Space Station Office,

in charge of Station development and design, and the Aeronautics and Space Technology and Commercial Programs. The latter offices plan the user programs as well as the selection and development of the payloads which go on board the Station.

Science payloads for the Space Station are among the payloads to be included on the STS and *Spacelab*. Among these payloads are both attached (to the Station itself) and pressurized projects. Some of the equipment is purely experimental in nature including that planned for cardiovascular experiments, radiation monitoring experiments, the Cosmic Ray Nuclei Experiment, and experiments in crystal growth and fluid dynamics.

This full-scale mock-up of a station module was built by Martin Marietta at the NASA Michould Assembly Facility in New Orleans, Louisiana.

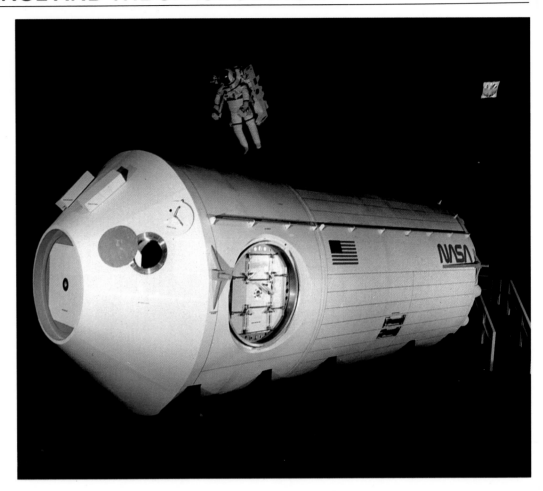

The Space Station will be the first long-term opportunity for work with animals, plants, and experimental equipment in weightlessness. NASA plans to utilize this opportunity by developing a module which will support studies of alterations in physiological functions, such as calcium excretion in plants, and studying how both human and non-human organisms function during long periods of weightlessness. This experimental module will also study the use of artificial gravity as a countermeasure to the effects of weightlessness on biological development. The first phase of the Station will house systems for data collection and animal development studies as well as the testing of new equipment.

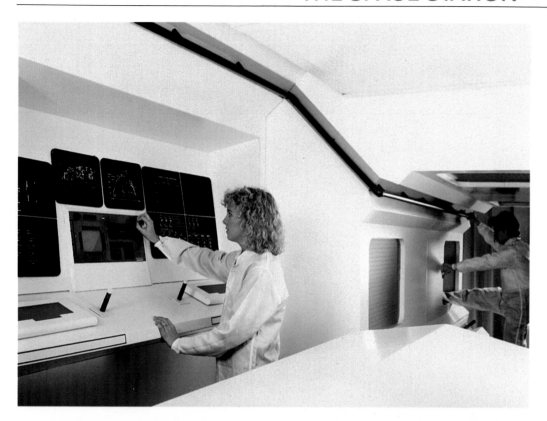

This McDonnell Douglas mock-up shows one of the habitat modules planned for the Station. Called the "quiet module" crew members would both sleep and work here.

A Rockwell International mock-up of the interior of a module.

# Life Sciences

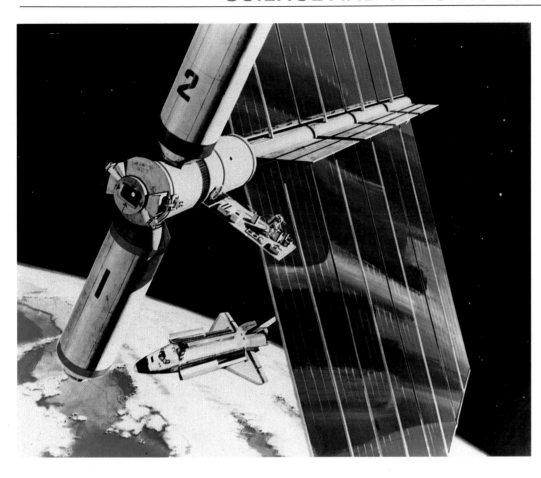

**A** space shuttle with a 200 kw power module and an experiment module.

A major research field likely to benefit from the Space Station is that of the life sciences. In this field, the overall objectives are to:

Determine the effects of prolonged space flight on humans and understand the basic physiological mechanisms of those effects.

Develop an effective means of maintaining spacecraft crews in good health, treating illnesses and injuries that occur in space, and permitting the crew members to reach the pinnacle of work productivity of which they are capable.

Learn how non-human organisms sense and react to varied levels of gravitational force.

Illustrate how life began here on Earth and what its distribution throughout the cosmos may be.

Characterize the role of life in processes that effect the Earth environment on a global scale.

Because of its responsibility for medical support to manned space operations, the life sciences project is directly concerned with biomedical problems in the Space Station and STS. With the current information already obtained, NASA anticipates that "adequate health maintenance measures can be available for Space Station operations when they begin." As space experimentation and Earth-based research grow throughout the 1980s, a healthy environment, tolerable living conditions, and medical treatment will no longer be hindrances to prolonged journeys into space.

Biomedicine and biology will be intertwined through mutual techniques, equipment, and experimental objectives. Both sciences will have important payload activities on the Space Station, and both must be gradually built up in stages from a single laboratory module for the Station's initial period, to a multimodule lab in years to come. Because of this, the life sciences project will plan its payload development on an integrated basis rather than on specific individual activities. A long-term goal of the life science experimentation is to develop practical biological methods for regenerating food, air, and water in life-support systems for missions to the Moon and to Mars.

Despite the fact that no specific module has been set aside for these endeavors, the capabilities built into the United States laboratory allow for extensive research in the life sciences. Space reserved for the United States in the Japanese and the European modules could also be allocated for the life sciences. Due to the outcry from academic and NASA scientists, research activities in the United States laboratory will emphasize materials research. Life science activities considered compatible will also, however, operate within this module. This is also expected to be the case with the European and Japanese modules. Where life sciences work is not compatible, it will be conducted else-

where. The current proposal is that the United States would receive a fraction of the space in the ESA module and the Japanese Experiment Module (JEM) for life science experiments, in exchange for the ESA and JEM receipt of resources from the United States laboratory module and a portion of the other two modules. This design has replaced the earlier "functional allocation" approach where each of the three modules would house specific functions.

# Earth Sciences

Many earth science experiments and studies are now being conducted on *Spacelab;* after a period of time, however, they will be modified to be performed on or near the Space Station. These experiments will depend upon the Station for servicing, assembly, and manned interactions. At this time, plans include a major facility, two principal investigator instruments, and an experimental program. NASA plans to utilize Space Station platforms of the Advanced X-Ray Astrophysics Facility and the Shuttle Infrared Telescope Facility. The Advanced Solar Observatory will be operational after the Solar Optical Tele-

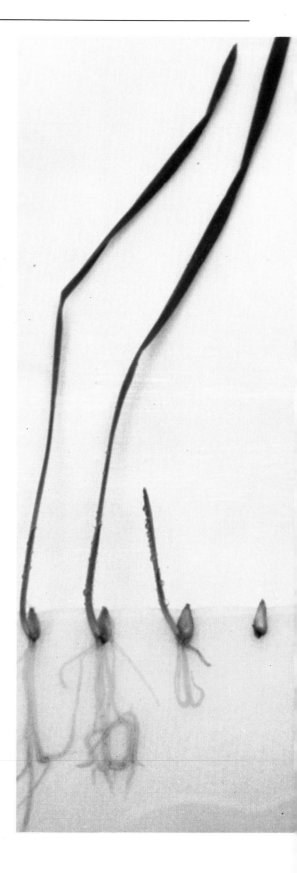

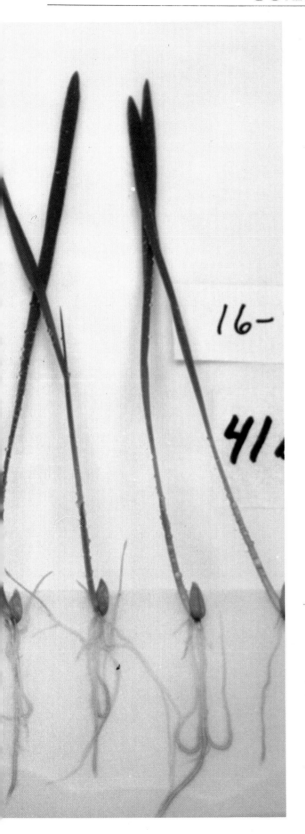

**P**ost flight photos of pine seedlings flown on *Spacelab 2.* The miniature greenhouses, called Plant Growth Units, allow investigators to monitor the effect of weightlessness on plant growth and the formation of lignin, a woody substance found in plants that permits them to grow upward against the pull of gravity.

**S**ome of the 3,000 bees onboard the shuttle *Challenger* are seen in this 35mm close-up of the aluminum box in which the young colony remained for the seven day mission.

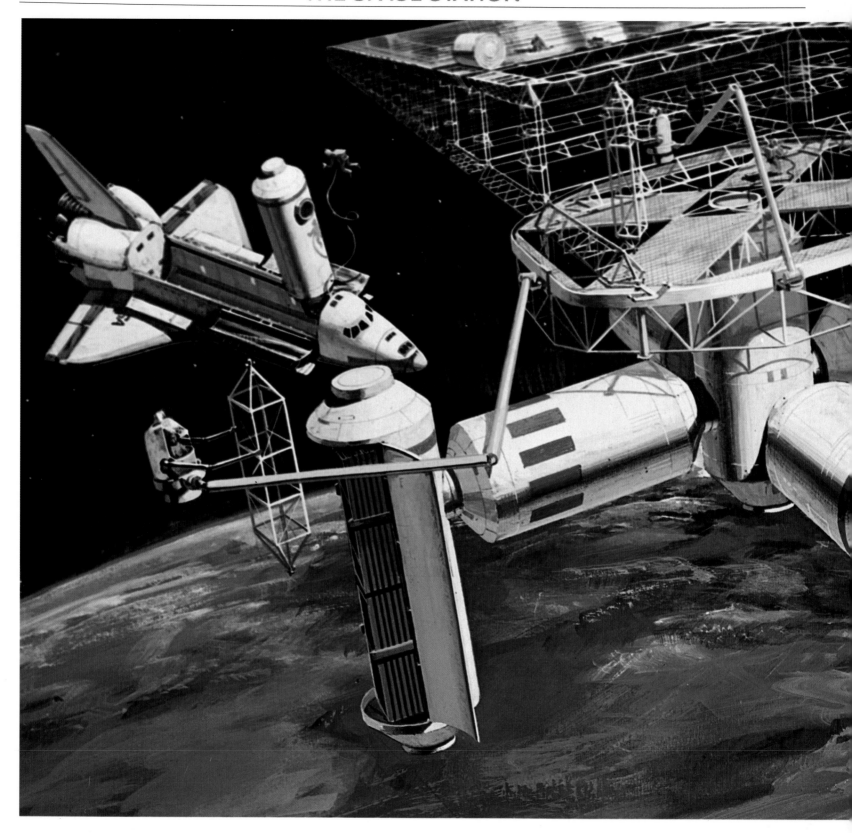

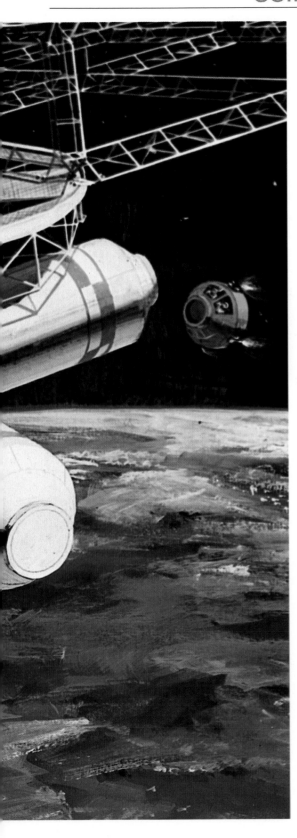

**T**his artist's rendering depicts activity at a possible manned station in Earth orbit.

scope and the Pinhole Occulator Facility are assembled on the Space Station or another, tethered (connected via cord) long-duration platform. The observatory will utilize high-resolution instruments to study all aspects of the Sun and plans are for it to advance solar physics drastically from an observational science to a science of prediction.

In the 1990s, the Earth Studies program will research long-term physical, biological, and chemical trends and alterations in the Earth's environment, including the atmosphere, lithosphere, magnetosphere, land masses, and oceans. By measuring the chemical cycles of nutrients, the effects of natural and human activities (such as pollution) on the Earth's environment will be better understood as models for predicting future effects. The space-based measurements will be performed from the polar platform and other free-flying platforms.

# Communications

Today, we are in an era appropriately called the "Information Age." Information is power and, as such, the building block for economic growth. News and data crisscross the globe, both as raw material and as a finished product. All of this information activity was made possible by communications satellites, which now make up a crucial and rapidly expanding portion of the communications industry. Between 1981 and the year 2000, estimates of the global market for communications satellite hardware alone total in excess of $38 billion.

In the field of communications, NASA program plans for the next decade include the development of technology to alleviate growing congestion in orbit and frequency allocation shortage, as well as the evolution of communications technology and systems that permit innovative new services in communications, navigation, and search and rescue, and finally, the support of its and the United States' interests in international and domestic communications regulations. Beyond the mid-1990s, world communications will probably progress toward centers of interconnection, like phone centers, using fiber optic cables for linking these high-density communications. Business communications, data distribution, and low-density communi-

cations would be distributed via satellite with access provided by intersatellite links. Therefore, the area of space below low Earth orbit and geostationary orbit will become a hot commodity for new and innovative communications services. Twenty-four years ago NASA launched the first communications satellite, today nearly thirty are in orbit.

Eventually the Space Station will become a complex of space systems in low Earth orbit incorporating co-orbiting platforms, tethered platforms, and spacecrafts free-flying platforms, an un-

**A** drawing of the interior mechanics of a beam builder that would construct components for a proposed space solar power system in geosynchronous orbit.

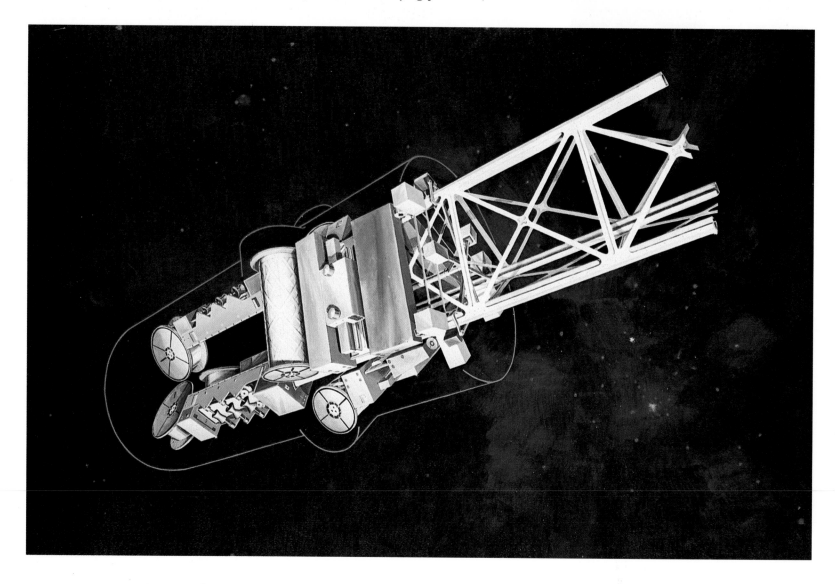

manned polar orbiting platform, and a manned module with research modules attached. This complex will, by necessity, be serviced by systems like the Shuttle and will access information by tracking data relay satellites and then communicating back with either these same satellites, the Shuttle, or the manned modules. The polar platform will be capable of "seeing" the Earth not visible to the manned modules and will conduct scientific experiments of the Earth with side-looking radars and imaging facilities.

The baseline communications system for the Space Station is the Tracking and Data Relay Satellite System. This system will handle the Station's needs by using laser and millimeter-wave links to assist the program in developing technology capable of supplying great amounts of data. Millimeter-wave communications at 60 gigahertz or higher will compete with laser communications. Both systems provide wide-band communications augmented by the vacuum of space, and each is among the most challenging procedures for making high data rate communications available between two low Earth orbits, below Earth orbit, and to the geosynchronous (one that shares the same orbit as Earth) orbit platform.

The Space Station will develop and test communications networks and technologies that can then be handed over to its operational systems. The laser communications system could, for example, be tested between the Advanced Communications Technology Satellite (ACTS) and the Shuttle, as a basis for upgrading the capability of the Tracking and Data Relay Satellite System, to handle the high (more than a gigabit) data rates planned for the 1990s.

NASA also plans to test a Spartan-

This painting by Denise Watts shows the initial stage of a mission to an earth-approaching asteroid. An orbital construction platform in permanent orbit provides power and a supply depot.

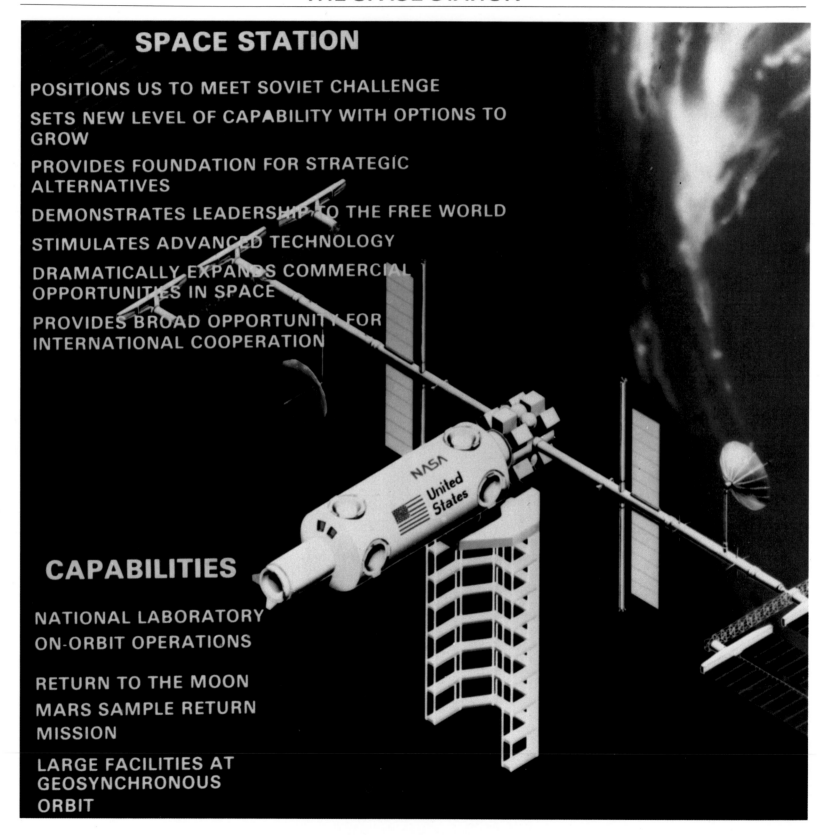

**SPACE STATION**

POSITIONS US TO MEET SOVIET CHALLENGE

SETS NEW LEVEL OF CAPABILITY WITH OPTIONS TO GROW

PROVIDES FOUNDATION FOR STRATEGIC ALTERNATIVES

DEMONSTRATES LEADERSHIP TO THE FREE WORLD

STIMULATES ADVANCED TECHNOLOGY

DRAMATICALLY EXPANDS COMMERCIAL OPPORTUNITIES IN SPACE

PROVIDES BROAD OPPORTUNITY FOR INTERNATIONAL COOPERATION

**CAPABILITIES**

NATIONAL LABORATORY
ON-ORBIT OPERATIONS

RETURN TO THE MOON
MARS SAMPLE RETURN
MISSION

LARGE FACILITIES AT
GEOSYNCHRONOUS
ORBIT

**R**ight: Studies of geostationary platforms include uses to aid in education, medical and health care, data and information exchange systems, and many other Earth resource monitoring measurements.

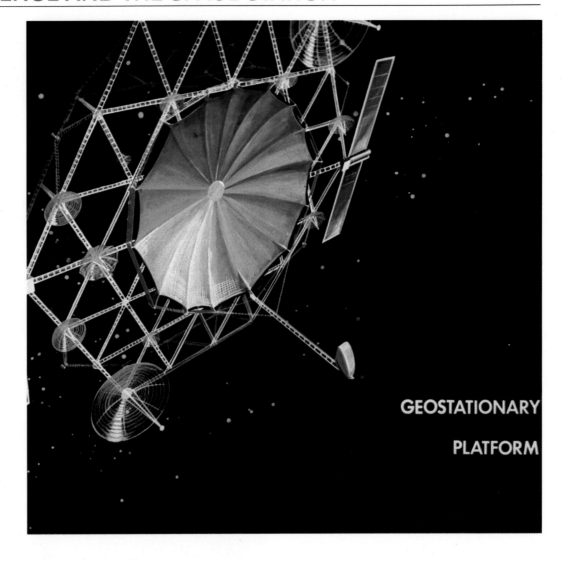

GEOSTATIONARY

PLATFORM

**O**verlaying this early space station design is a listing of the capabilities and broad benefits it would provide.

type maneuverable free-flyer (see page 112) that could be instrumented with experimental communications technology and could be controlled from the Space Station. This spacecraft could make a series of experimental communication link-ups with the Shuttle, the ACTS, or with the Earth. Its experimental uses could include: serving as a test vehicle for an Antenna Far Field Test Range by traversing through space to test the side lobes of an antenna attached to the Space Station; testing the operation of related payloads; and transmitting the collected data to the Shuttle

or the Space Station for quick analysis.

Large antenna development is currently being managed by NASA with the Harris Corporation, under contract to Langley Research Center, in Virginia, for a 50-foot (15-meter) antenna, and for a 180-foot (55-meter) antenna developed by Lockheed Missiles and Space Company, who are under contract to the Jet Propulsion Laboratory in Texas.

NASA believes that by the year 2000, the Space Station could include a geosynchronous platform (see page 113) and orbital transfer vehicles operating in both low Earth orbit and geosynchro-

This particular model of a space station features a modular space platform of rotating pallets containing space science and applications payloads; airlocks join the three manned modules.

nous orbit. Far more advanced than the innovative concept of 1977, (see page 56) it would present new challenges in architecture. It might be composed either of multiple clusters of small antennas or of giant antennas, or both. It may have central or distributed direct power systems instead of self-contained modular power systems. It could even be held together, without a rigid structure, but with electromagnetic radiation.

Presently, NASA has a Joint Endeavor Agreement with the private sector to test a special low Earth orbit satellite capable of receiving, on command, data from a ground position over which it is passing and then, also on command, transmitting, or "dumping," the data to another ground point. This "store and dump" satellite would have various applications, would be low cost, and could be launched from the Shuttle and maintained by the Space Station.

As NASA learns how to best utilize all of the Space Station's capabilities, its basic method of conducting research and development on antennas will certainly change, and construction, tests, and subsequent antenna launchings to other orbits will become commonplace.

Presently, concepts for the use of laser intersatellite links yielding space communications are being studied at NASA. One plan is for a commercial data relay satellite to use laser links to access satellites in the geostationary arc, low Earth orbit, geostationary orbit, and even to access high-flying aircraft. NASA theorizes that potential customers would be marketing services such as crop analysis or mapping and exploration for Earth resources, such as water, oil, or gems. In the end, NASA may provide the leadership in developing technologies and systems that interconnect the planet with all its satellites.

# Moving to the Future

As the twentieth century draws to a close, the frontiers of space call to us like a beacon. Most of the programs and concepts outlined in the previous chapters of this book will reach fruition with the end of this century. The start of a new century is slightly more than a decade away and preliminary planning of possible systems, activities, and programs for the beginning of the twenty-first century are

**A**bove: Three astronauts perform EVA's on a station in low-Earth orbit. Left: This is a Boeing Aerospace company design: Note the habitation windows.

**P**revious page: Shown here (not to scale) is an illustration by the Marshall Space Flight Center that incorporates stations in both high inclination and low inclination orbits.

already under way at NASA. While goals change with each accomplishment, and detailed plans would be premature, a wide range of programs and alternatives must be initiated to set new goals and objectives and to assist in the identification of programs and technologies that would provide the greatest potential. In 1987, looking toward the twenty-first century, NASA originated the Office of Explorations which will conduct detailed studies of possible human exploration of the Moon, Mars, and other celestial bodies.

If the United States is to be the leader in improving the general well-being of humankind on Earth and in space beyond the twentieth century, long-range goals must be initiated for advanced scientific knowledge, space exploration, Earth applications, and commercial uses. Ongoing space studies, research, and development will also be important to initiate and to develop even more innovative systems and techniques than those developed at the end of this century. At NASA, concepts already exist for manned and automated space missions to achieve these goals and to provide unprecedented scientific and technical benefits. These missions will utilize all avenues of accessible space—low Earth orbit, higher-energy Earth orbit, lunar orbit, the lunar landscape, and the environs of the inner planets—in the beginning of the twenty-first century.

These twenty-first century accomplishments in space exploration, Earth applications, science, and commercial uses hinge on two trends. The first is that of the increasing capabilities of space systems including accessibility, time in space, payloads, and greater sophistication of operations. The second is the increase in capability of instrumentation for detection, resolution, accuracy, data collection, and improved power and cooling mechanisms. As these trends and the space infrastructure they produce develop, so too, will our future in space.

# A Lunar Base

In 1987, Dr. Sally K. Ride, the first American woman to fly in space, was appointed to head a panel to determine what NASA's goals beyond the Space Station should entail. In a sixty-three-page report to the NASA administrator, the Ride panel suggested that "America should not rush headlong towards Mars; we should adopt a strategy to continue an orderly expansion outward from the Earth." Advising against a spectacular "race to Mars," the Ride panel recommended a study of Earth from satellites and unmanned spacecraft and, most importantly, the establishment of a permanent scientific outpost on the Moon. By mining the Moon's resources, the United States would begin learning how to live both on and off of the lunar land,

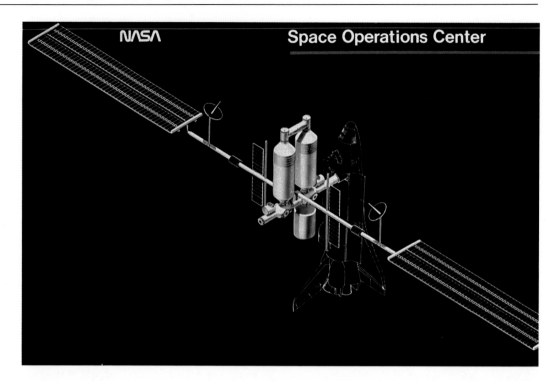

The safety of both the Space Operations Center and its crew is a major design consideration in any Station. Thus, while the Station would usually be manned, it could also be operated automatically.

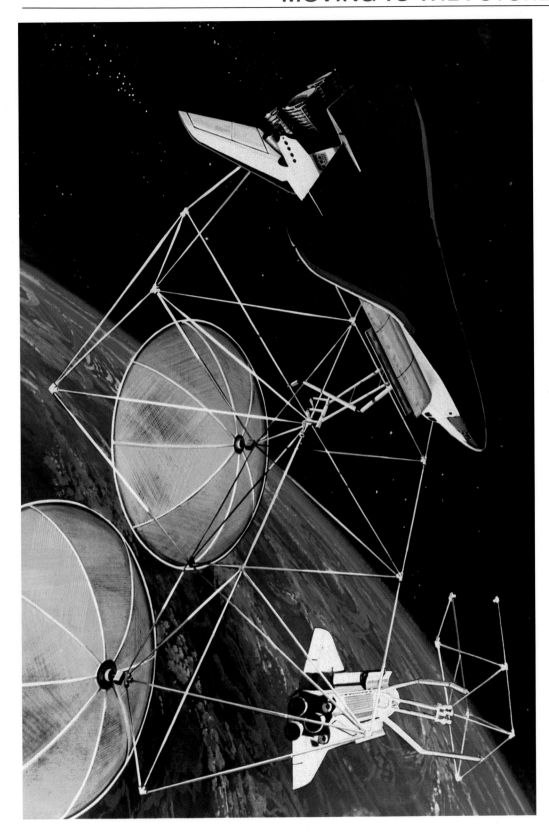

which would then prepare NASA for a trip to Mars.

When NASA eventually begins regular voyages to the Moon, the routine access to its surface will enable the first systematic study of another body in space. Intensive sample collection and scientific treks conducted by humans, and instrument vehicles guided and installed by humans, will assist in determining the details of the lunar structure, composition, and history. Soil studies can also be conducted to record solar and cosmic ray particle penetration. The lunar surface will be made available as a platform for astronomical observations. At the outset, NASA will conduct studies with small prototype, human-tended instruments and will concentrate on sky surveys and the detection of unpredictable phenomena, such as supernovas. Because a lunar outpost needs technological research, it is anticipated that initially the systems will be derived from the Space Station and will be adapted to withstand the environment of the Moon, while the functioning of the base will evolve from only casual human tending to a permanently manned outpost.

**U**tilizing opportunities that exist only in space, the shuttle would assist in the construction of satellites.

**A**n artist's conception of a 1990s mission armada on Mars. Here, two manned capsules have already landed.

# Mars

Much of the preliminary exploration work on Mars must be done with unmanned vehicles. NASA plans a Mars sample return mission in the early 1990s, which will require new technologies. The strategy includes deployment of payload elements by the Shuttle; in-orbit assembly of the full Mars craft; deployment of an orbiter into a Mars trajectory; sample collection, launch, orbital rendezvous, and docking at the Space Station; and the transfer of samples to the Shuttle for return to Earth.

The Mars Observer, to be launched in 1990, will study Mars for two years and then answer questions regarding climate, atmosphere, geochemistry, and other tests necessary before NASA can send a human mission to the Red Planet.

While the Ride panel cautioned against making Mars the primary goal of space exploration, it believed Mars to be the "ultimate objective" of the next forty years. While the report made no estimate of costs, it did state that a mission to deploy humans to Mars by the year 2005 would require a tripling of NASA's budget. Much needs to be learned about the mining of alien surfaces and the effects of long-term flight before a manned expedition to Mars will be feasible.

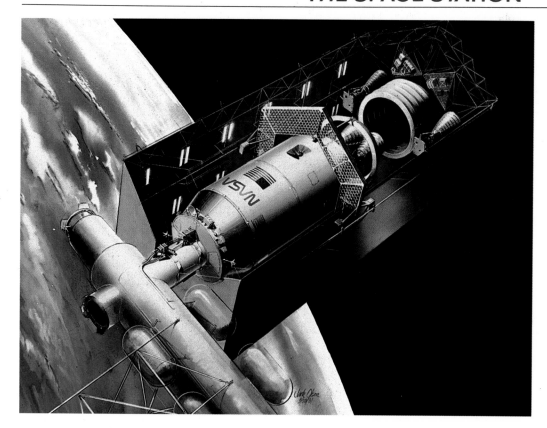

This drawing (left) shows the Shuttle and Station positioned in low-Earth orbit. To the left, an upper stage propels a satellite into geosynchronous orbit.

# Low Orbit Facilities

As discussed previously, NASA has a joint endeavor agreement with the private sector to construct an experimental communications satellite. This satellite would have various applications, one being the transmission of data from space buoys to a data collection facility.

This particular system represents a strong movement toward non-utilization of geostationary arc orbit locations for communications using the low Earth orbit paths. As the Space Station develops free-flying and other assorted platforms, low Earth orbit systems for communications will come into being using such tools as fiber optics and lasers.

# Planetary Probes

In its attempt to understand the world around us, NASA has created one of the most exciting and important scientific activities of this century. In this attempt, NASA has led the way in the exploration of our solar system and has brought knowledge, prestige, and pride to all Americans. United States spacecraft were the first to reach Mercury, Venus, and Mars. Only United States spacecraft have traversed the asteroid belt into the outer solar system—four visiting Jupiter and three going to Saturn. *Voyager 2* passed by Uranus in 1986 and will encounter Neptune in 1989. *Pioneer 10* is already in interstellar space, its distance from the sun is greater than

that of any known planet. During this formative period, NASA has followed the National Research Council's Space Science Board recommendation for a balanced approach to exploration; that it should "move forward on a broad front to all accessible planetary bodies beginning with reconnaissance, into exploration of selected planets and lastly, to study a limited number of cases."

The Board's Committee on Planetary and Lunar Exploration has divided the bodies in the solar system into three categories: the inner planets (Mercury, Venus, Earth, the Moon, and Mars), the small bodies (asteroids, comets, and meteorites), and the outer planets (Jupiter and its satellites, Saturn and its satellite Titan, Uranus, Neptune, and Pluto). The recommendation of the

This drawing (left) shows the Shuttle and Station positioned in low-Earth orbit. To the left, an upper stage propels a satellite into geosynchronous orbit.

Artist Don Davis painted this depiction (right) of *Voyager 2* as it looks back upon the planet Neptune and its moon, Triton, seven hours after its closest approach to the planet, scheduled for August 24, 1989.

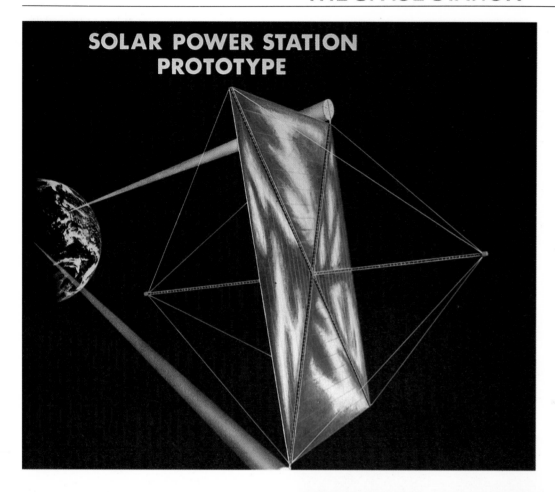

## SOLAR POWER STATION PROTOTYPE

A preliminary configuration of a solar power station.

Jerry L. Ross (wearing EMU with red stripes) at work with EASE (right).

A Shuttle orbiter and space station docking.

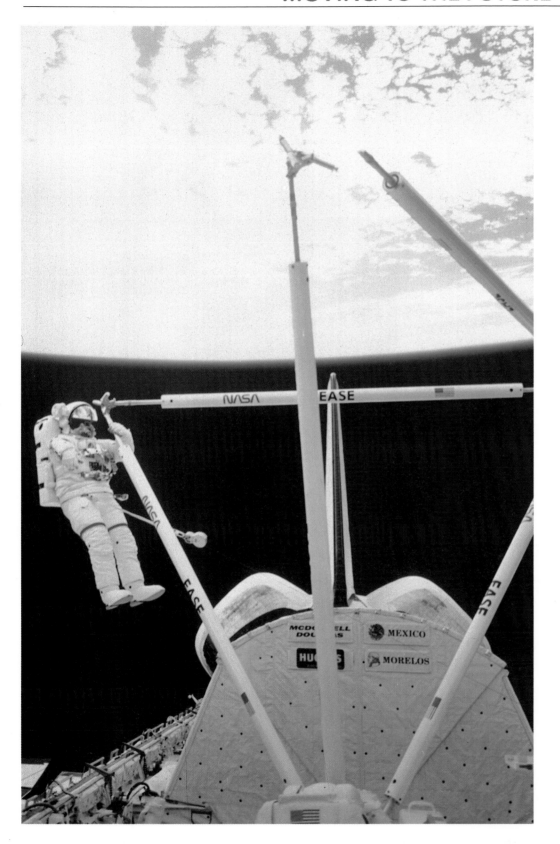

committee was to concentrate during the 1980s on the inner planets (Venus, Earth, and Mars), and much has already been learned from the probes.

The outer planets offer a different motivation for study—understanding the formation and evolution of our solar system. *Voyager 1* and *Voyager 2* have both completed studies of Jupiter and Saturn, and *Voyager 2* is enroute first to Uranus and then to Neptune, completing the initial phase of exploration. The secondary stage will launch *Galileo*, a major probe and orbiter, to investigate Jupiter in the late 1980s.

*Galileo's* mission is to study the chemical composition and structure of Jupiter's atmosphere, its physical state, and the surface composition of its moons. *Galileo* will also access the structure, composition, and dynamics of the magnetosphere of Jupiter. NASA expects *Galileo* to answer questions posed by the encounters of *Voyager 1* and *2*, which found the magnetic field surrounding Jupiter to drastically change in size and measurement from visit to visit, as well as probing other complex structural occurrences within the atmosphere.

*Galileo*, originally scheduled for launch in 1986, is to be orbited before 1990. The spacecraft consists of an orbiter and an atmospheric probe and uses the Shuttle and a Centaur upper stage to set it on a trajectory to Jupiter. During *Galileo's* twenty-month, eleven-orbit tour, it will, at least once, fly through and map Jupiter's magnetic tail in an area not previously analysed.

Scheduled for 1989 is the launch of a Venus Radar Mapper, which will greatly contribute to the understanding of that planet and will determine such things as whether there was once water on its surface and how its evolution compared to that of the Earth.

# CONCLUSION

Today, the Soviet Union, Europe, Japan, and China have spacecraft and launch vehicles of proven reliability and complexity because they understand that exploitation of space is an unspoken imperative. The response of the United States should not be that of a bitter competitor but, rather, that of a proud people dreaming the impossible dream, on a direct course with destiny. Despite its enormous potential and considerable practicality, the Space Station may represent something more far-reaching—the symbolic commitment of the United States to space exploration and the prosperous and intellectually stimulating future that exploration promises.

An artist's conception of the permanently manned Space Station currently being developed by NASA. In the foreground of the picture is the solar dynamic power system, which uses concentrated light from the sun to heat a fluid, which turns a generator, to provide electrical power for the Station. Solar array panels are also used to generate electricity for the station.

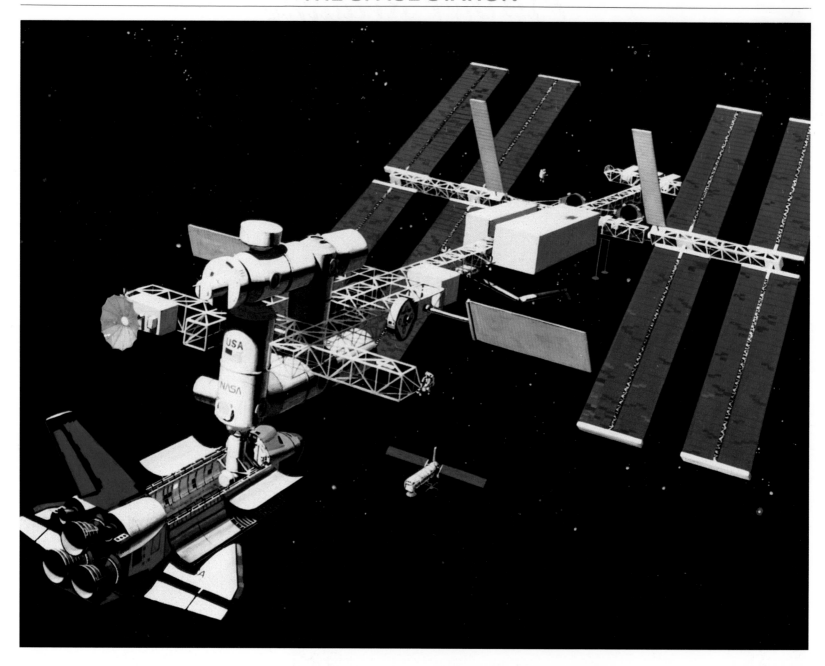

**A**bove: depicted here is a high angle view of the "power tower," with a space platform and an earth resources package. This picture was produced by the Rockwell International Corporation of Downey, California. Right: This concept developed by TRW of Redondo Beach, California, shows a space shuttle orbiter undergoing servicing at a space station. Near the orbiter are the rotating pallets capable of holding space science and applications payloads.

# Further Reading

Adelman, Saul J., *Bound for the Stars*. Englewood Cliffs, NJ: Prentice-Hall, 1980.

Baker, Wendy. *NASA: America in Space*. NY: Crescent Books, 1986.

*The Rocket: the History and Development of Rocket and Missile Technology*. NY: Crown Publishers, 1978.

Stine, G. Harry. *The Third Industrial Revolution*. New York: Putnam, 1975.

Tanner, Dan and George Johnson. *Cities in Space*. Eugene, OR: Harvest House, 1979.

Von Braun, Wernher, and Frederick I. Ordway. *History of Rocketry and Space Travel*. NY: Crowell, 1975.

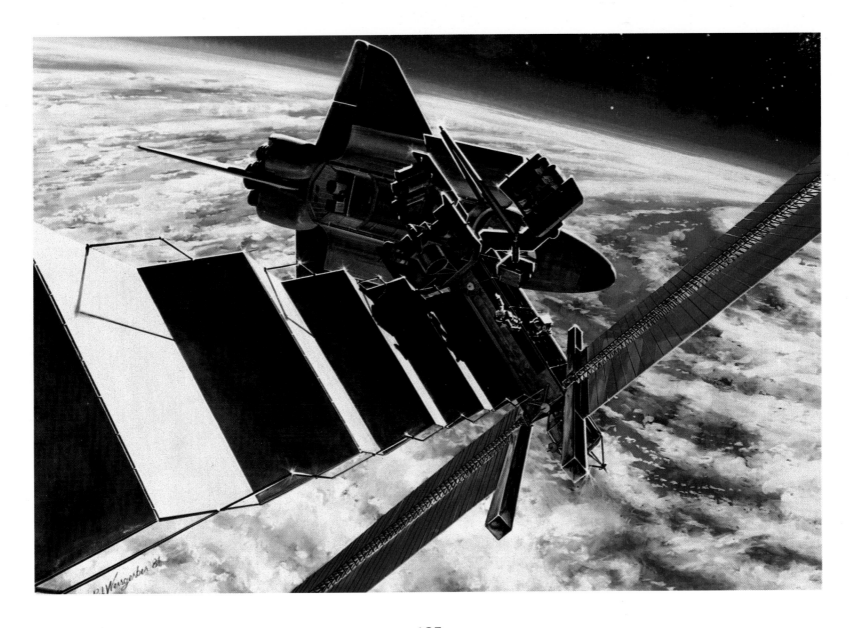

S hown here is a space station illustration by TRW's Space and Technology Group. Portrayed is a later-phase space station consisting of dual solar array sets, and logistics, command, and habitat modules. This station would be tended by the Space Shuttle.

# *Index*

# N

# O

# P

# R

# S

Pictured here is a Space Shuttle Orbiter Solar Array System—a solar panel to receive the sun's rays and possibly add twenty-two days life to a Shuttle mission. This system would be deployed by the Shuttle's remote manipulator arm.